Applied Landscape Ecology

Applied Landscape Ecology

Francisco Castro Rego
Professor of Landscape Ecology
Centre for Applied Ecology "Prof. Baeta Neves"
InBio and School of Agriculture
University of Lisbon
Lisbon, Portugal

Stephen C. Bunting
Emeritus Professor of Landscape and Rangeland Ecology
College of Natural Resources
University of Idaho
Moscow, Idaho, USA

Eva Kristina Strand
Professor of Landscape Ecology
College of Natural Resources
University of Idaho
Moscow, Idaho, USA

Paulo Godinho-Ferreira
Senior Researcher of Landscape Ecology
Centre for Applied Ecology "Prof. Baeta Neves"
InBio and Institute for Agrarian and Veterinarian Research
Lisbon, Portugal

The right of Francisco Castro Rego, Stephen C. Bunting, Eva Kristina Strand and Paulo Godinho-Ferreira to be identified as the authors of this work has been asserted in accordance with law.

Registered Office(s)
John Wiley & Sons, Inc., 111 River Street, Hoboken, NJ 07030, USA
John Wiley & Sons Ltd, The Atrium, Southern Gate, Chichester, West Sussex, PO19 8SQ, UK

Editorial Office
The Atrium, Southern Gate, Chichester, West Sussex, PO19 8SQ, UK

For details of our global editorial offices, customer services, and more information about Wiley products visit us at www.wiley.com.

Wiley also publishes its books in a variety of electronic formats and by print-on-demand. Some content that appears in standard print versions of this book may not be available in other formats.

Library of Congress Cataloging-in-Publication Data

Names: Rego, Francisco Castro, author. | Bunting, Stephen C., author. |
    Strand, Eva K., author. | Godinho-Ferreira, Paulo, author.
Title: Applied landscape ecology / by Francisco Castro Rego, Stephen C.
    Bunting, Eva Kristina Strand, Paulo Godinho-Ferreira.
Description: First edition. | Hoboken, NJ : John Wiley & Sons, 2018. |
    Includes index. |
Identifiers: LCCN 2018015504 (print) | LCCN 2018035715 (ebook) | ISBN
    9781119368243 (pdf) | ISBN 9781119368229 (epub) | ISBN 9781119368205
    (cloth)
Subjects: LCSH: Landscape ecology.
Classification: LCC QH541.15.L35 (ebook) | LCC QH541.15.L35 R45 2018 (print)
    | DDC 577–dc23
LC record available at https://lccn.loc.gov/2018015504

Cover Design: Wiley
Cover Image: © Luisa Nunes

Set in 10/12pt Warnock by SPi Global, Pondicherry, India
Printed and bound in Singapore by Markono Print Media Pte Ltd

10 9 8 7 6 5 4 3 2 1

Dedicated to our students, stewards of future landscapes

Contents

Foreword

As urbanization spreads swiftly and visibly across the land and climate change promises further severe effects, increasingly landscape ecology is needed at the core of solutions. Building from the basic patch–corridor–matrix model and an array of spatial principles, the field describes, analyzes, and prescribes patterns for science and society. Not surprisingly, landscape ecology has strengthened key dimensions of forestry, conservation biology, wildlife management, landscape architecture, transportation, and urban planning.

Along with principles and applications, theory has grown, as noted by Princeton ecologist, Simon Levin, who participated in landscape ecology's 1983 North American launching: "…landscape ecology has developed into one of the most vibrant branches of ecological science, with exceptionally strong links between theory and practice. Indeed, it is hard to think of any area in ecology where theory has had a greater impact on application, or where applications have done more to stimulate creative theory." (Foreword to Barrett, G.W., Barrett, T.L., and Wu, J. (eds) (2015) *History of Landscape Ecology in the United States*, Springer.)

Several valuable landscape ecology books are used in teaching and applications. Yet students, professors, and practitioners alike would welcome a succinct, well-illustrated handbook of core principles, plus equations, highlighting how to measure and analyze patterns that affect processes. This is now in your hand.

Global in perspective, the book builds on a tradition of joint North American and European co-authors (Forman books with, respectively, French (*Landscape Ecology*, John Wiley & Sons, 1986), Dutch (*Changing Landscapes: An Ecological Perspective*, Springer, 1990), and Norwegian (*Landscape Ecology Principles in Landscape Architecture and Land-Use Planning*, 2nd edn, Island Press, 1996) co-authors). Embellishing the pages of this book are examples from islands and arid lands to forests and agriculture, even cities.

The authors begin with a somewhat novel organizing theme, points, lines, and patches. For each, simple familiar patterns, formulas, examples, and analyses are presented, and gradually build to slightly more complex principles. The approach is amazingly logical and lucid. The reader explores the core of landscape ecology and grasps the essence of concepts. Subsequent chapters highlight the vertical dimension, movement through landscapes, composition, configuration, and landscape dynamics, using the earlier basics in progressively more complex analyses. Each of these later chapters takes an extra step or two, opening up useful frontiers. The final chapter applies the principles and tools, along with the transition matrix, to landscape planning and management.

This book emerges from researchers with years of teaching landscape ecology. While reading, I felt an author was in the room with me talking through concepts, and showing how to measure and analyze the diverse patterns outside. The delightfully clear writing and explanations made some previously muddy ideas crystallize, eureka-like.

Imagine having the following for both teaching and learning: abundant illustrations, often in color, of the place, pattern, and/or species considered; many key landscape ecologists briefly introduced with a photograph; frequent GIS vector, or especially clear raster, representations and analyses; patterns often compared, using equations and empirical data; classic, as well as new, field studies analyzed; metrics for detecting pattern and interpreting process; how to calculate common indices; full references given as Endnotes at the end of each chapter where cited; and every chapter with succinctly stated key points.

Frontiers for landscape ecology abound. Land patterns are crucial for towns/villages and their surroundings. The big picture is still needed for cities/suburbs (e.g., *Urban Ecology: Science of Cities*, Cambridge, 2014), urban regions (*Urban Regions: Ecology and Planning Beyond the City*, Cambridge, 2008), and transportation (*Handbook of Road Ecology*, John Wiley & Sons, 2015). Solutions on the border of vision and feasibility, such as for transportation (*Solutions*, 2, 10–23, 2011) and best places for the next billion people (*Nature*, 537, 608–611, 2016; *Handbook on Biodiversity and Ecosystem Services in Impact Assessment*, Edward Elgar, 2016), build on landscape ecology. Economists and other social scientists, who relish equations and horizontal axes representing time, could richly add to this field. Landscape ecology books with, or for, planners could quickly change planning and improve our land. Opportunities await.

Turning the pages ahead reveals a unique book. All landscape ecologists and related colleagues will value, and should have, these lucid explanations of core principles using analytic methods. This volume is valuable for academics teaching a course, researchers face-to-face with landscape patterns, and perceptive practitioners improving our land.

Richard T.T. Forman
Harvard University

Preface

The formative stage of Landscape Ecology included many of the principles that are a part of today's ecological science but was primarily descriptive in nature. As with most subdisciplines within ecology, landscape ecology has become increasingly more quantitative. Increased analytical power via geographic information systems (GIS) and the availability of broad-scale remotely-sensed data has made this possible. Development of statistical methods that consider the location of objects and their spatial dependency has further advanced our ability to quantify landscape patterns and processes. Experience has shown that many of today's issues related to natural resources require a broad-scale, and perhaps long-term, perspective to adequately address. These include issues such as invasive species, water quality, fire management, sustainable development, and species conservation. This broad-scale perspective nearly always includes lands managed by different land stewards with varying land management objectives. Clearly landscape ecology must include quantitative characteristics to communicate between land stewards, governmental agencies, and the general public. In addition, quantitative metrics are necessary to describe the efficacy of landscape management decisions and actions.

We have selected the organism as our focus from which we will analyze our landscapes. We believe that the organism-centered approach has many advantages. The organism-oriented focus can be justified by: (1) most people can readily identify with the individual organism and the species, (2) our current global concern related to biological diversity is focused on species, and (3) we often measure the effects of ecological processes such as disturbance, succession, competition and climatic and global change in terms of their effects on species. The individual organism then becomes the basic unit of consideration, which can be aggregated into populations, populations can be aggregated into communities living in patches, and patches can be aggregated into landscapes. Greater and greater numbers of patches can be aggregated to create larger and larger landscapes.

The authors have all taught landscape ecology for many years in their respective universities at the undergraduate and/or graduate level. We have also cooperated in various research projects related to landscape ecology in Europe and the United States in forested, rangeland, and agricultural dominated landscapes. The need for this book arose from our teaching landscape ecology courses. The experiences on research projects have also indicated a need to have a new comprehensive text related to landscape metrics in order to communicate with graduate students and others involved in the research. We also believe that the approach taken in this book, used already and tested successfully in classes in both the University of Idaho and of Lisbon, provides a step-by-step development of landscape metrics that is best suited for use as a course textbook in landscape ecology.

The organization of the book begins with the simplest situation, the description of single elements (points, linear features, and patches) and then proceeds to describe more complex landscape metrics incorporating the concept of time. While the description of one patch is not landscape analysis, per se, these patch metrics are the basis from which many of the subsequent analysis methods of the landscape are derived and a thorough understanding of them is required for successful landscape-scale analysis. All analysis methods are illustrated with simple examples that the reader will be able to derive using only a spreadsheet or a hand calculator. This will enable the reader to understand more fully the derivation of the metrics as well as their advantages and limitations. The application of these metrics is then elaborated with more complex examples from the published literature.

We would like to thank the many colleagues and former students over the past decade with whom we have discussed the concepts of landscape ecology, landscape metrics, and the approaches of the presentation of these concepts and metrics in a classroom setting. We would particularly like to recognize the conversations with Penny Morgan, Wendell Hann, Steve Petersen, Rick Miller, Steve Knick, Marta Rocha, and Ana Abrantes. We have benefitted greatly from these discussions and are indebted to all these people. Finally, we would like to thank Liliana Bento and also Miguel Geraldes for his diligence and hard work in helping us in the final stages of the book's preparation. We are not sure that we could have done it without you. Keep looking up, Miguel.

FCR, SCB, EKS, & PG-F

Figure The four authors of the book, from left to right: Paulo Godinho-Ferreira, Stephen C. Bunting, Francisco Castro Rego, and Eva Kristina Strand.

Authors

Francisco Castro Rego
Professor of Landscape Ecology
Centre for Applied Ecology "Prof. Baeta Neves"
InBio and School of Agriculture, University of Lisbon
Lisbon, Portugal

Stephen C. Bunting
Emeritus Professor of Landscape and Rangeland Ecology
College of Natural Resources
University of Idaho
Moscow, Idaho, USA

Eva Kristina Strand
Professor of Landscape Ecology
College of Natural Resources
University of Idaho
Moscow, Idaho, USA

Paulo Godinho-Ferreira
Senior Researcher of Landscape Ecology
Centre for Applied Ecology "Prof. Baeta Neves"
InBio and Institute for Agrarian and Veterinarian Research
Lisbon, Portugal

1

Concepts and Approaches in Landscape Ecology

1.1 The Historical Development of Landscape Ecology as a Science

Ecology as a written science probably has its known beginnings in ancient Greece with Aristotle and particularly with his successor, Theophrastus, who was one of the first philosophers to study "the relationships between the organisms and their environment". This definition of the term Ecology that was first used two millennia later by the German zoologist Ernst Haeckel, who, in 1866, associated the Greek words *Oikos* (house) and *Logos* (science) (Figure 1.1).

Haeckel further expanded the definition of Ecology in his writings in 1869[1]: "By ecology we mean the body of knowledge concerning the economy of nature, the investigation of the total relations of the animal both to its inorganic and to its organic environment; including above all, its friendly and inimical relations with those animals and plants with which it comes directly or indirectly into contact."

Other subdisciplines of ecology focus on the study of the distribution and abundance of individuals of the same species (population ecology)[2], on the interaction between populations (community ecology)[3], or, especially after the very influential book published in 1953 by Eugene Odum[4], on the study of ecosystems (systems ecology). Ecology has expanded from populations to communities and ecosystems, and more recently to landscape scales.

The English word "landscape" first appeared in the late sixteenth century when the term *landschap* was introduced by Dutch painters who used it to refer to paintings whose primary subject matter was natural scenery, associating the word "land" (of Germanic origin) and the suffix "schaft" or "scape", meaning shape[5] (Figure 1.2).

In the seventeenth and eighteenth centuries, "landscape" continued to be associated with paintings, but a new meaning of the term developed when Alexander von Humboldt (1769–1857) started the new science of plant geography. Humboldt explored the visual qualities of painted landscapes transforming the concept of landscape from its primary visual meaning into an abstract entity by finding its ecological unity[6] (Figure 1.3). The concept of landscape was moving from art to ecological science.

Following the work of Humboldt, it was another German geographer, Carl Troll, who first coined in 1939[7] the term "landscape ecology" hoping that a new science could be developed that integrated the spatial approach of geographers and the functional approach of ecologists (Figure 1.3).

Applied Landscape Ecology, First Edition. Francisco Castro Rego, Stephen C. Bunting,
Eva Kristina Strand and Paulo Godinho-Ferreira.
© 2019 John Wiley & Sons Ltd. Published 2019 by John Wiley & Sons Ltd.

Figure 1.1 The Greek philosopher Theophrastus (371–287 BC) (left) and the German ecologist Ernst Haeckel (1834–1919 AC) (right). *Source:* https://upload.wikimedia.org/wikipedia/commons/3/38/ Theophrastus._Line_engraving._Wellcome_V0005785.jpg, https://upload.wikimedia.org/wikipedia/ commons/2/2f/Ernst _Haeckel_2.jpg (3 December 2017).

However, the science of landscape ecology would be one of the latest forms of ecology to develop. It was not until the 1980s that the concept was more widely developed and works on landscape ecology started to be produced in Europe and in North America with the books by Vink (1983)[8], Naveh and Lieberman (1984)[9], and Forman and Godron (1986)[10]. After the publication of this latter book, which is now considered to be a main foundation of this science, references to landscape ecology started to become common in scientific literature and further developed after the beginning of the publication of the scientific journal of *Landscape Ecology* in 1987.

After that initial period, landscape ecology studies also became common in other basic and applied ecological journals. The development of landscape ecology was rapid and several important books were produced after the 1980s. In 1995 Forman published his comprehensive award-winning book *Land Mosaics: The Ecology of Landscapes and Regions*[11] (Figure 1.4).

Since 1995 several other books have been published by various authors in both Europe and North America, as in the Netherlands by Zonneveld (1995)[12], in Italy by Farina (1998)[13] and by Ingegnoli (2002)[14], in France by Burel and Baudry (1999)[15], in the United States by Turner, Gardner, and O′Neill (2001)[16], by Coulson and Tchakerian (2010)[17], and by Forman, again, with others, on *Road Ecology: Science and Solutions* (2003)[18] and on *Urban Ecology: Science of Cities* (2014)[19].

Also, many edited books with applications of landscape ecology analyses were published since the late 1980s, such as those by Turner (1987)[20], Turner and Gardner (1991)[21], Bissonette (1997)[22], Klopatek and Gardner (1999)[23], Sanderson and Harris (2000)[24], Wiens, Moss, Turner, and Mladenoff (2006)[25], Wu and Hobbs (2007)[26], McKenzie, Miller, and Falk (2011)[27], and Perera, Drew, and Johnson (2012)[28].

Figure 1.2 Landscape painting of Richmond castle (1639) by the Dutch landscape painter Alexander Keirincx (1600–1652). *Source*: Yale Center for British Art, Paul Mellon Collection: Netherlandish Painters Active in Britain in the Sixteenth and Seventeenth Centuries, http://ezine.codart.nl/17/issue/46/artikel/netherlandish-painters-active-in-britain-in-the-16th-and-17th-centuries/?id=191 (17 February 2017).

Figure 1.3 Painting of the German naturalist Alexander von Humboldt (1769–1859) (left) and photo of the German geographer Carl Troll (1899–1975) (right). *Source:* Portrait of Alexander von Humboldt by Friedrich Georg Weitsch, 1806, https://en.wikipedia.org/wiki/Alexander_von_Humboldt#/media/File: Alexandre_humboldt.jpg, https://de.wikipedia.org/wiki/Carl_Troll (17 February 2017).

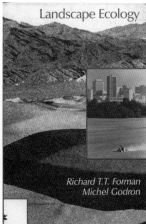
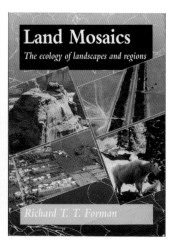

Figure 1.4 The USA scientist Richard Forman (born 1935) with two of his fundamental books in Landscape Ecology. *Source:* Harvard University, Graduate School of Design, http://www.gsd.harvard.edu/person/richard-t-t-forman/ (2 May 2017).

Figure 1.5 The top image is a LiDAR acquisition of hundreds of conifer trees. The lower left image shows the branching structure of a tree and the lower right image is a synoptic view of one complete 360 degree scan of a mixed conifer forest near Moscow, Idaho, USA. The images were taken with a Leica green terrestrial laser scanner. Colors correspond to the relative return intensity of the LiDAR instrument, with greens and yellows showing high intensity returns and red showing low intensity returns. *Source:* Courtesy of Lee Vierling and Jan Eitel, University of Idaho.

Landscape ecology was definitely settled as a new science and several books were published on the corresponding quantitative methods, as that in the Netherlands by Jongman, ter Brak, and van Tongeren (1995)[29] on data analysis in community and landscape ecology. A major development of analytical methods used in landscape ecology came in 1995 with the release of the FRAGSTATS software in association with the publication of a very useful USDA Forest Service General Technical Report by McGarigal and Marks[30]. Due to its popularity the program has been updated and recently upgraded to accommodate Arc-GIS10 (version 3.4) and it has been central to other books for quantifying and measuring landscape characteristics, such as that of Leitão, Miller, Ahern, and McGarigal (2006)[31]. From this perspective, it is also important to recognize the book edited by Gergel and Turner (2002)[32] covering many of the quantitative methods in landscape ecology.

During the last decades the science of landscape ecology developed with input from new technologies (Figure 1.5) and contributions from many other disciplines such as geology, soil science, plant ecology, wildlife ecology, conservation biology, genetics, human ecology, urban ecology, and landscape architecture. In addition, landscape ecology was enhanced by the rapid advancement of computer sciences, remote sensing, geographic information system technologies, and landscape modeling. Remote sensing images obtained from satellite platforms displaying features of the Earth's surface first became available in the latter half of the twentieth century. Finally, landscape ecologists were able to view landscape patterns over large land areas and it became increasingly feasible to quantify change in both spatial and temporal dimensions. Currently the landscape ecology perspective is essential in addressing broad-scale complex issues such as those associated with global change.

For example, the Landsat satellites have been collecting multispectral images at 15–80 m resolution since the 1970s at a 16-day interval and the moderate-resolution imaging spectroradiometers (MODISs) have been collecting 250–1000 m resolution images every 1–2 days since 2002. Technological advancement has also been made in data collection at very fine spatial scales. Light detection and ranging (LiDAR) is a technique with the ability to map objects in three dimensions by measuring the time it takes for a laser signal to travel from the sensor to the object and back to the sensor. Sensors mounted on unmanned aircraft systems (UASs) can provide images with subcentimeter resolutions.

Computer-based geographic information systems (GISs) have made it possible for landscape ecologists to handle the large amounts of spatial data available today. GIS data are constructed of spatial representations of points, lines, polygons, or pixels, accompanied by a database that describes each feature and its location. Increasing power of computational systems continues to allow us to analyze datasets at larger extents and with finer resolution.

1.2   Hierarchical Levels in Ecology

The hierarchy of ecological units and related sciences has been well represented as a set of Chinese boxes, each fitting inside the next larger box (Figure 1.6).

This hierarchy of disciplines shows that science can develop in two opposite directions: downwards to the genetic and molecular level and upwards from the organism to the population, community, habitat/ecosystem, and landscape (total human ecosystem) level.

The combination of the different levels to understand ecological processes and spatial patterns is fundamental in Landscape Ecology. The combination of the two more extreme levels (genes and landscapes) has been the basis of the recent development of the new discipline of Landscape Genetics.

The emphasis on the genetic levels has been made possible by the fast development of modern equipment, methods, and techniques. Also, the emphasis on genes has been encouraged by the very challenging views of authors such as Richard Dawkins, who, in his 1976 book *The Selfish Gene*[34] expresses his "stranger than fiction" feelings about the "astonishing truth" that "we are survival machines – robot vehicles blindly programmed to preserve the selfish molecules known as genes".

The combination of Landscape and Genetics leading to the development of Landscape Genetics occurred after 2003 following the work of Manel and others[35] with the aim of understanding how landscape features influence genetic variation, an approach that has the advantage of not requiring discrete populations to be identified in advance. However, it requires knowledge of genetics and methods using molecular markers or genomic scans that are beyond the scope of this book.

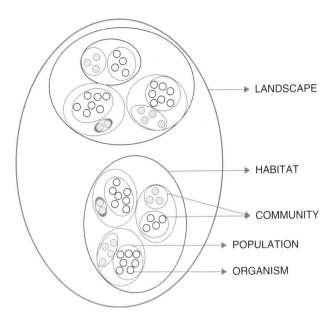

LANDSCAPE

HABITAT

COMMUNITY

POPULATION

ORGANISM

Figure 1.6 The ecological hierarchy and its scientific disciplines through five levels of integration (adapted from Koestler by Naveh and Liebermann, 1994)[33].

Landscape Genetics has evolved and has stimulated research in various areas, as in the study of the effect of landscapes on evolutionary processes[36]. The main emphasis of landscape genetics has nevertheless continued to be on the processes and patterns of gene flow[37], that is, how genes are incorporated into the gene pool of one population from other populations. This involves the detection of genetic variation and the analysis of its relation to landscape characteristics, as in the existence of barriers or of habitats that facilitate gene flow.

As the landscape characteristics that influence gene flow are the same as those that influence the movement of organisms from one habitat patch to another through the landscape, the same landscape metrics (habitat fragmentation or connectivity) apply to both gene flow and organism movements.

Therefore, for the approach and the structure of this book we have selected the organism as our focus from which we will analyze landscapes. The organism-centered approach has an advantage in that most people can readily identify with the individual organism and its life-sustaining requirements.

The individual organism, the basic unit of consideration, can be aggregated into populations (individuals of the same species living in the same area), populations can be aggregated into communities (populations of different species living in the same patches), and patches (representing different habitats) can be aggregated into landscapes. Greater and greater numbers of patches can be aggregated to create larger and larger landscapes.

1.3 The Spatial Hierarchy of Land

Similar to the ecological hierarchy, there is also a spatial hierarchy of land. Forman illustrates this hierarchy well: "Suppose you had a giant zoom lens hooked up to your

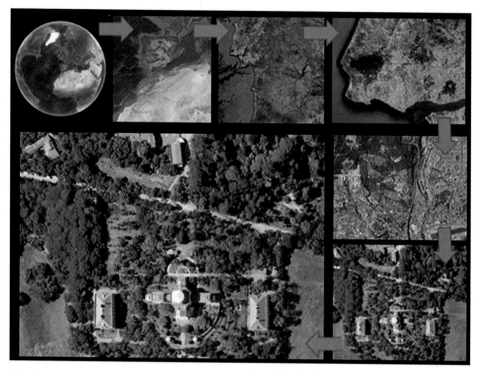

Figure 1.7 The spatial hierarchy of land from the planet to the Iberian Peninsula and northern Africa, to the central Portugal region, to the city of Lisbon, to a mosaic of urban and green patches, and finally to the individual buildings and trees around the Astronomical Observatory of Ajuda in two scales. Viewing altitudes range from 5×10^7 m for the planet to 5×10^2 m with all the intermediate images in a sequence of powers of ten, 10^7 m to 10^3 m. *Source:* Google Maps.

personal spaceship. You begin with a view of the whole planet, and slowly and evenly close the lens until you have a microscopic view of soil particles. At any point you would probably see a mosaic, a heterogeneous pattern of patches and corridors"[38] (Figure 1.7).

We can now better define, in this hierarchy of land, the relationship between concepts such as ecosystems and landscapes. Ecosystems, as the concept was initially described by Tansley[39], and landscapes have the common characteristic of not having a predetermined spatial scale. The scale of each can be set by the user. However, ecosystem ecology, as developed by Odum[40], focuses on the effects of ecological processes such as nutrient cycling, the hydrologic cycle, and energy flow, and their influence on the relationships among plants, animals, air, water, and soil within a spatial unit. Landscape ecology is unique in its focus on the relationships among spatial units. Thus, landscapes must be heterogeneous with respect to at least one type of element.

Following these concepts it is now possible to define landscape as "a heterogeneous land area composed of a mosaic of interacting spatial units". According to this definition there is not a specific size for a landscape or a specific scale for the analysis but they should be meaningful to the particular organism or process of interest. Also, according to the objective of the analysis, the various spatial units present in the landscape can be represented by different landscape elements, such as patches, lines, or points.

1.4   Fundamental Concepts: Landscape Scale and Size, Pattern, Process, and Change

A frequently asked question is "How does the landscape ecology approach differ from that of the other types of ecology? What are its fundamental concepts?" This is difficult to answer precisely. Many scientists consider the broad spatial and temporal scales that are often studied by landscape ecologists as an important distinction. However, landscapes need not be defined as broad scale and nor are broad scales the sole dominion of landscape ecologists.

Certainly those ecologists who study the biological response to long-term paleoclimatic change or primary ecological succession have a long-term perspective. Many other ecologists also study ecosystems with very broad spatial scales. In addition, landscape ecological approaches may be applied to small areas over short time periods, such as movements of small mammals within a landscape mosaic.

Perhaps the distinguishing feature of landscapes, regardless of their spatial or temporal scale, is that they are, by nearly all definitions, composed of mosaics of interacting patches of different types. The arrangement of these patches affects the nature of their interactions and the arrangement changes with time, as do the interactions (Figure 1.8).

Other ecological disciplines attempt to minimize differences between sampled patches and develop stratified sampling strategies to account for and minimize these differences. Landscape ecology has emphasized the inclusion of the heterogeneity that is present.

From the perspective of an organism, we may define landscape as the area containing a mosaic of patches that are meaningful to its perception of the surrounding environment.

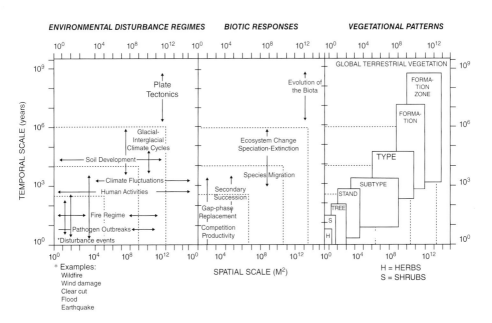

Figure 1.8 Time and space scales of disturbances, biotic responses, and vegetational patterns as summarized by Delcourt and Delcourt[41].

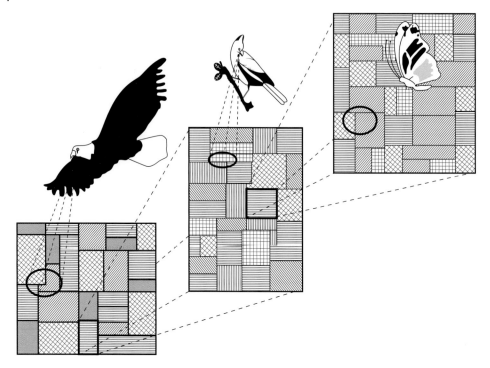

Figure 1.9 Changing perception of the environment depending on organism scale[44].

As a consequence the extent (or size) of a landscape (overall area encompassed by its boundary) differs among organisms, generally between its normal home range and its regional distribution[42].

The importance of considering the various perceptions of the environment is well illustrated in the work of McGarigal and Marks[43]: "Because the eagle, cardinal and butterfly perceive their environments differently and at different scales, what constitutes a single habitat patch for the eagle may constitute an entire landscape for the cardinal, and a single habitat patch for the cardinal may comprise an entire landscape for the butterfly that perceives patches on an even finer scale" (Figure 1.9).

Also, the recognition of pattern depends on the size of the individual units of observation (grain) used in the representation of the landscape, which is in turn generally associated with scale. As in photography, the ability to recognize pattern is also related to contrast, or the amount of difference between adjacent elements and the relative abruptness of their boundary[45]. The recognition of pattern is therefore related to grain and contrast but the perception of the whole depends more on the interrelations of the elements than on their details.

A simple illustration of the process of pattern recognition is demonstrated by the comparison of satellite images of western North America (Figure 1.10). It is not necessary to have a very fine grain image to start detecting the patterns. In the image in the middle the western coast is already recognized without full knowledge of details of each pixel but simply by the perception of the whole.

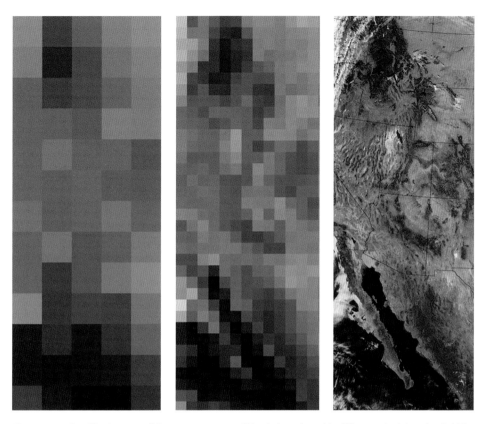

Figure 1.10 Satellite Images of the western coast of North America with different pixel sizes (grain). The general pattern can already be recognized in the image with intermediate grain. *Source:* Google Maps.

Similar illustrations are presented by Naveh and Lieberman[46] as a demonstration of the importance of the holistic systems approach. They suggested that the comprehension of the landscape is better achieved by the identification of the structural relationships of the components than by their detailed study. This represents the advantage of landscape ecology over community and systems ecology in the recognition of pattern.

Landscape ecology is also an attractive approach to both the study and management of natural and human-dominated ecosystems for a number of reasons. The inclusion of a mosaic of interacting patches presents a more realistic setting. It is certainly well known that patches in a mosaic often interact[47].

In forest management, for example, this knowledge has been used for generations when planning timber harvest units where natural regeneration from seeds produced in surrounding uncut areas (patches) was expected. For this process the temporal scale is set by the harvesting age and the spatial scale by the seed dispersion distance (Figure 1.11).

However, if initially focusing on plants and animals in more natural environments, the concept of a landscape also typically includes human activity[48]. Humans

Figure 1.11 Forest pattern resulting from wood harvesting in Western Oregon, USA. *Source:* Marli Miller, geologypics.com, http://www.marlimillerphoto.com/images/Res-58lr.jpg.

have influenced all ecosystems on the planet and in agricultural and urban landscapes human activity is a dominant factor in determining the mosaic expressed on the land.

As previously mentioned, there is no specific temporal or spatial scale at which landscape ecology focuses, as the appropriate scale varies with the objective of the analysis. From the perspective of humans, a landscape may be a kilometer-wide area and patches can then be approximately 1–10 ha in natural vegetation, but may be larger or smaller in human-dominated landscapes such as urban or agricultural landscapes (Figure 1.12). As for temporal scales, humans tend to view short-term processes as occurring within a small fraction of human lifespans and long-term processes as occurring in multiple human lifespans.

In a very influential paper, Monica Turner[49] referred to landscape ecology as the study of the reciprocal influences of pattern (structure) on process (function) and change (dynamics) of those interactions through time. She further indicated that landscape structure should be quantified to understand the relationships of pattern with ecological processes, as in the understanding of the influence of spatial heterogeneity in the spread of disturbances or the flows of energy, matter, nutrients, and organisms.

The assessment of landscape structure, function, and change through time is intended to make it possible to understand the dynamic relationships between pattern and process. However, as Monica Turner[49] rightly pointed out, as landscapes are spatially heterogeneous areas, the measurement of spatial pattern and heterogeneity is dependent upon the scale at which the measurements are made and the structure, function, and change of

Figure 1.12 For many humans the size of the perceived landscape was for a long time limited by borders or walls. The walls of the city of Bragança, Portugal. *Source:* F. Pratas, 2007, https://commons.wikimedia.org/w/index.php?curid=2443851.

landscapes are themselves scale-dependent, and the scale at which studies are conducted may profoundly influence the conclusions.

1.5   The Representation of the Landscape and its Elements

From the beginning of cartography map features have been described by points, lines, and polygons (Figure 1.13). Cartographers have used points to identify the location of small islands or cities, lines to illustrate river courses or roads, or polygons for bigger islands or continents.

With aerial photography and satellite imagery, grid-based or raster data provided new ways to represent the same landscapes in very different ways (Figure 1.14).

However, in many cases, these raster maps derived from aerial photographs or satellite images have to be transformed to vector maps for simplicity of analysis. In spite of the increasing use of aerial photography and satellite imagery, much of the spatial analysis continues to be done with points, lines, and polygons at the various scales, with geographic information systems using the vector digital representation of landscapes in two-dimensional (2D) maps (Figure 1.15). Also the third (vertical) dimension can be captured in 2D vector maps by using elevation contour lines (Figure 1.16).

In the vector representation of landscapes in maps, points are often used to represent the location of small objects (at the selected scale). Points can be associated with individual trees, buildings, and water holes, while lines can be used to look at coastlines, rivers or road networks, and polygons to represent patches with different land cover types, such as agricultural fields, urban areas, forests, shrublands, grasslands, or sand beaches. The elements of the landscape can therefore be spatially represented in a map by these three different landscape elements (points, lines, or/and patches), as illustrated in Figure 1.17. The detection of the pattern of a landscape can be assessed by the detection of the pattern of the different landscape elements (Figure 1.18).

Figure 1.13 Nicolas Desliens Map of the World (1566). *Source:* Nicolas_Desliens1.JPG, https://commons.wikimedia.org/wiki/File:Nicolas_Desliens_Map(1566).jpg#/media/File:Nicolas_Desliens_Map_(1566).jpg (17 February 2017).

Figure 1.14 Land cover patterns in the Northern Great Plains, South Dakota, USA. *Source:* Google Maps.

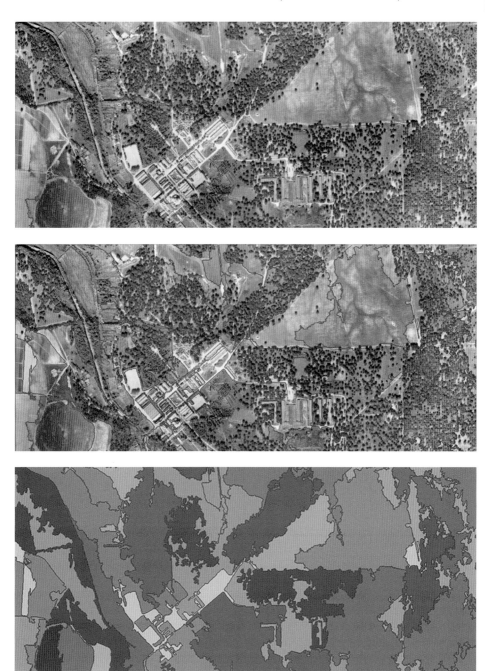

Figure 1.15 The automatic transformation of a digital aerial photograph of an area with forests (cork oak), agriculture, urban development, and roads in Rio Frio, close to Lisbon, into a vector map of land cover classes using the eCognition software, showing the original image (upper), the automatic delineation of polygons (center), and the resulting vector map with different classes (lower). *Source:* Unpublished report by Cadima and Rego for the Management Plan of the Herdade de Rio Frio (Portugal).

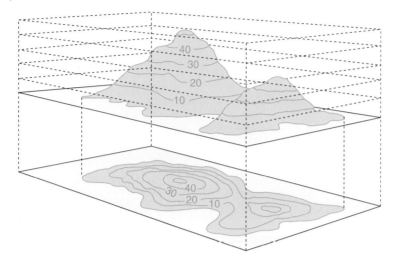

Figure 1.16 Contour lines represent 3D objects in 2D maps. *Source:* Gemma Nelson, https://www.ordnancesurvey.co.uk/blog/2015/11/map-reading-skills-making-sense-of-contour-lines/.

Figure 1.17 Aerial image of Porto do Carro, Portugal, showing a mosaic including some scattered trees and isolated houses (points), river and road networks (lines), and forest, agricultural, and urban areas (patches). *Source:* Google Earth, 2015.

We can therefore recognize from the scale of the mosaic the following spatial units representing landscape elements:

Point – representing a significant landscape feature whose geographical location is important but whose area is not relevant for the organism or the landscape process of interest. In fact, points do not exist, but they can represent small patches for the purpose of the analysis.

Line – representing a significant landscape feature that differs from the adjacent land on both sides, whose location and linear dimension are important for the functions of conduit (ecological line corridor), filter, or barrier but whose width is not relevant

Figure 1.18 The pattern of landscape elements viewed from an airplane can give "landscape detectives" extensive information about the ecology of a landscape. *Source:* Karen Bloomberg, http://10steps.sg/inspirations/photography/11-types-of-beautiful-photos-from-plane-window/ (17 February 2017).

for the organism (does not provide habitat) or the process of interest. In fact, lines do not exist, but if a patch is long and its width is small at the scale and for the purpose of the analysis, lines can be an adequate representation of the object. However, if width is small but relevant the representation should be a strip corridor, which would be represented with a patch.

Patch – representing a relatively homogeneous nonlinear area that differs from its surroundings. Differences between patches may be based on ecological (i.e. species composition) or other differences (i.e. land ownership, different watersheds) (Figure 1.18).

Forman[50] further suggests the possibility of the recognition of another landscape element, the matrix or the background type (or class) in a mosaic, characterized by "extensive cover, high connectivity, and/or major control over dynamics".

The same approach has been extended to professions of a different nature. Rapidly these approaches were demonstrated to be useful in Landscape Architecture and Land-Use Planning[51], and works in the urban context creating a new branch of this science: the Urban Landscape Ecology (Figure 1.19).

However, Landscape Ecology has also been expanding in its applications outside terrestrial ecosystems. Landscape Ecology has recently been used as a theoretical and analytical frameworks for "evaluating the ecological consequences of spatial patterns and structural changes in the submerged landscapes of coastal ecosystems"[53]. This expansion was the beginning of a new branch of Landscape Ecology: Seascape Ecology.

URBAN LANDSCAPE ECOLOGY
SCIENCE, POLICY AND PRACTICE

Edited by ROBERT A. FRANCIS, JAMES D. A. MILLINGTON
and MICHAEL A. CHADWICK

earthscan
from Routledge

Figure 1.19 A book on Urban Landscape Ecology[52] resulting from a Conference of the International Association for Landscape Ecology in the United Kingdom in 2014 with the participation of academics, practitioners and policy makers from around the world. *Source:* Routledge, https://www.routledge.com/info/contact.

Similarly to terrestrial landscape elements, the concepts of patches, matrix, and mosaics can also be defined and analyzed in submerged landscapes (Figure 1.20).

The branch of Seascape Ecology has developed in various contexts (Figure 1.21). In Portugal a recent study used this approach to evaluate the characteristics of the underwater landscape related to SCUBA diving activity in the Algarve[55]. The development of this new scientific branch has recently resulted in a book written in 2017 by a pioneer of this new branch of science with the appealing title: *Seascape Ecology: Taking Landscape Ecology into the Sea*[56].

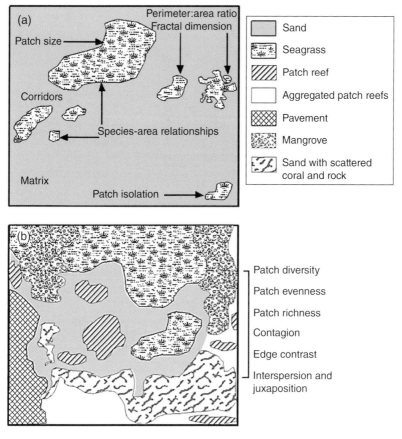

Figure 1.20 The seascape can be represented as (A) above: a patch-matrix model with seagrass patches surrounded by a matrix of sediment or (B) below: a patch-mosaic model, more complex, with different patch types originating more complex metrics[54]. *Source:* Wedding, L.M., Lepczyk, C.A., Pittman, S.J., *et al.* (2011) Quantifying seascape structure: extending terrestrial spatial pattern metrics to the marine realm. *Marine Ecology Progress Series*, 427, 219–232.

Figure 1.21 Seascape Ecology was the subject of the Annual Conference of the International Association of Landscape Ecology UK in 2015. *Source:* wikipedia.org, https://en.wikipedia.org/wiki/Underwater (3 December 2017).

Regardless of the nature of the landscape (urban or rural, terrestrial or submerged) there are common features in detecting the pattern of the landscape and its elements. This allows the current book to be applied to the various types of landscapes to be analyzed as they can be represented by elements such as points, lines, or patches that are changing in time due to landscape processes.

In this book we follow a sequential approach starting from the analysis of pattern of groups of points to groups of lines and then patches, presenting some possible interpretations associating the detected patterns of processes before discussing methods of analysis of change.

Key Points

- Landscape can be defined as a heterogeneous land area composed of a mosaic of inter-acting spatial units, or mosaics of interacting patches of different types.
- Landscape ecology is the study of the reciprocal influences of pattern (structure) on process (function) and change (dynamics) of those interactions through time. Landscape structure should be quantified to understand the relationships of pattern with ecological processes.
- Landscape analysis does not have a specific size or scale. Size and scale should be meaningful to a particular organism or process of interest.
- Temporal or spatial scale at which landscape ecology focuses varies with the objective of the analysis.
- The extent (or size) of a landscape differs among organisms or process of interest, and the recognition of pattern depends on the grain (or size of the individual units of observation) used in the representation of the landscape.
- The measurement of spatial pattern and heterogeneity is dependent upon the scale at which the measurements are made and the structure, function, and dynamics of landscapes are themselves scale-dependent; the scale at which studies are conducted may profoundly influence the conclusions.
- A landscape can be spatially represented in a map by different landscape elements: points, lines, and/or patches.

Endnotes

1 Allee, W.C. (1949) *Principles of Animal Ecology*, W.B. Saunders, Philadelphia, PA.
2 Andrewartha, H.G. and Birch, C. (1954) *The Distribution and Abundance of Animals*, University of Chicago Press, Chicago, IL.
3 Daubenmire, R. (1968) *Plant Communities. A Textbook of Plant Synecology*, Harper and Row, New York, NY.
4 Odum, E.P. (1953) *Fundamentals of Ecology*, W. B. Saunders Company, Philadelphia, PA.
5 Makhzoumi, J. and Pungetti G. (1999) *Ecological Landscape Design and Planning*, Spon Routledge, New York, NY.
6 Kwa, C. (2005) Alexander von Humboldt's invention of the natural landscape. *The European Legacy: Toward New Paradigms*, 10, 149–162.

7 Troll, C. (1939) *Luftbildplan und Ökologische Bodenforschung*, Zeitschraft der Gesellschaft fur Erdkunde Zu, Berlin.

8 Vink, A.P.A. (1983) *Landscape Ecology and Land Use*, Longman Press, New York, NY.

9 Naveh, Z. and Lieberman, A.S. (1984) *Landscape Ecology – Theory and Application*, Springer-Verlag, New York, NY.

10 Forman, R.T.T. and Godron, M. (1986) *Landscape Ecology*, John Wiley & Sons, New York, NY.

11 Forman, R.T.T. (1995) *Land Mosaics: The Ecology of Landscapes and Regions*, Cambridge University Press, Cambridge, MA.

12 Zonneveld, I.S. (1995) *Landscape Ecology*, SPB Academic Publishing, Amsterdam.

13 Farina, A. (1998) *Principles and Methods in Landscape Ecology*, Chapman and Hall, New York, NY.

14 Ingegnoli, V. (2002) *Landscape Ecology: A Widening Foundation*, Springer, Berlin.

15 Burel, F. and Baudry, J. (1999) *Écologie du Paysage. Concepts, Methods et Applications*, Editions TEC & DOC, Paris.

16 Turner, M.G., Gardner, R.H., and O'Neill, R.V. (2001) *Landscape Ecology in Theory and Practice: Pattern and Process*, Springer, New York, NY.

17 Coulson, R.N. and Tchakerian, M.D. (2010) *Basic Landscape Ecology*, KEL Partners, College Station, TX.

18 Forman, R.T.T, Sperling, D., Bissonette, J.A., *et al.* (2003) *Road Ecology: Science and Solutions*, Island Press, Washington, DC.

19 Forman, R.T.T. (2014) *Urban Ecology: Science of Cities*, Cambridge University Press, Cambridge, MA.

20 Turner, M.G. (ed.) (1987) *Landscape Heterogeneity and Disturbance*, Ecological Studies 64, Springer-Verlag. New York, NY.

21 Turner, M.G. and Gardner, R.H. (eds) (1991) *Quantitative Methods in Landscape Ecology: The Analysis and Interpretation of Landscape Heterogeneity*, Springer-Verlag, New York, NY.

22 Bissonette, J.A. (ed) (1997) *Wildlife and Landscape Ecology: Effects of Pattern and Scale*, Springer Verlag, New York, NY.

23 Klopatek, J.M. and Gardner, R. (eds) (1999) *Landscape Ecological Analysis: Issues and Applications*, Springer-Verlag, New York, NY.

24 Sanderson, J.G. and Harris, L.D. (2000) *Landscape Ecology: A Top-down Approach*, Lewis Publishers, Boca Raton, FL.

25 Wiens, J.A., Moss, M.R., Turner, M.G., and Mladenoff, D.J. (eds) (2006) *Foundation Papers in Landscape Ecology*, Columbia University Press, New York, NY.

26 Wu, J. and Hobbs, R.J. (eds) (2007) *Key Topics in Landscape Ecology*, Cambridge University Press, Cambridge, UK.

27 McKenzie, D., Miller, C., and Falk, D.A. (eds) (2011) *Landscape Ecology of Fire*, Springer, New York, NY.

28 Perera, A.H., Drew, C.A., Johnson, C.J. (eds) (2012) *Expert Knowledge and Its Application in Landscape Ecology*, Springer, New York, NY.

29 Jongman, R.H.G., ter Brak, C.J.F., and van Tongeren, O.F.R. (eds) (1995) *Data Analysis in Community and Landscape Ecology*, Cambridge University Press, Cambridge, UK.

30 McGarigal, K. and Marks, B.J. (1995) *FRAGSTATS: Spatial Pattern Analysis Program for Quantifying Landscape Structure*, USDA Forest Service General Technical Report PNW-GTR-351, Pacific Northwest Research Station, Portland, OR.

31 Leitão, A.B., Miller, J., Ahern, J., and McGarigal, K. (2006) *Measuring Landscapes: A Professional Planner's Manual*, Island Press, Washington, DC.

32 Gergel, S.E. and Turner, M.G. (eds) (2002) *Learning Landscape Ecology: A Practical Guide to Concepts and Techniques*, Springer, New York, NY.

33 Naveh, Z. and Lieberman, A.S. (1994) *Landscape Ecology: Theory and Application*, Springer-Verlag, New York, NY.

34 Dawkins, R. (1978) *The Selfish Gene*, Oxford University Press, New York, NY.

35 Manel, S., Schwartz, M., Luikart, G., and Taberlet, P. (2003) Landscape genetics: combining landscape ecology and population genetics. *Trends in Ecology and Evolution*, 18 (4): 189–197.

36 Manel, S. and Holderegger, R. (2013) Review: Ten years of landscape genetics. *Trends in Ecology and Evolution*, 28, 614–621.

37 Holderegger, R. and Wagner, H. (2008) Landscape genetics. *BioScience*, 58, 199–207.

38 Forman, R.T.T. (1995) *Land Mosaics: The Ecology of Landscapes and Regions*, Cambridge University Press, Cambridge, MA, p. 11.

39 Tansley, A.G. (1935) The use and abuse of vegetational concepts and terms. *Ecology*, 16, 284–307.

40 Odum, E.P. (1953) *Fundamentals of Ecology*, W. B. Saunders Company, Philadelphia, PA.

41 Delcourt, H.R. and Delcourt, P.A. (1988) Quaternary landscape ecology: Relevant scales in space and time. *Landscape Ecology* 2, 23–44.

42 McGarigal, K. and Marks, B.J. (1995) *FRAGSTATS: Spatial Pattern Analysis Program for Quantifying Landscape Structure*, USDA Forest Service General Technical Report PNW-GTR-351, Pacific Northwest Research Station, Portland, OR.

43 McGarigal, K. and Marks, B.J. (1995) *FRAGSTATS: Spatial Pattern Analysis Program for Quantifying Landscape Structure*, USDA Forest Service General Technical Report PNW-GTR-351, Pacific Northwest Research Station, Portland, OR, p. 4.

44 McGarigal, K. and Marks, B.J. (1995) *FRAGSTATS: Spatial Pattern Analysis Program for Quantifying Landscape Structure*, USDA Forest Service General Technical Report PNW-GTR-351, Pacific Northwest Research Station, Portland, OR.

45 Forman, R.T.T. (1995) *Land Mosaics: The Ecology of Landscapes and Regions*, Cambridge University Press, Cambridge, MA.

46 Naveh, Z. and Lieberman, A.S. (1994) *Landscape Ecology: Theory and Application*, Springer-Verlag, New York, NY.

47 McGarigal, K. and Marks, B.J. (1995) *FRAGSTATS: Spatial Pattern Analysis Program for Quantifying Landscape Structure*, USDA Forest Service General Technical Report PNW-GTR-351, Pacific Northwest Research Station, Portland, OR.

48 Naveh, Z. and Lieberman, A.S. (1994) *Landscape Ecology: Theory and Application*, Springer-Verlag, New York, NY.

49 Turner, M. (1989) Landscape ecology: the effect of pattern on process. *Annual Review of Ecology and Systematics*, 20, 171–197.

50 Forman, R.T.T. (1995) *Land Mosaics: The Ecology of Landscapes and Regions*, Cambridge University Press, Cambridge, MA.

51 Dramstad, W., Olson, J.D., and Forman, R.T.T. (1996) *Landscape Ecology Principles in Landscape Architecture and Land-Use Planning*, Harvard University Graduate School of Design, American Society of Landscape Architects, and Island Press, Washington, DC.

52 Francis, R.A., Millington, J.D.A., and Chadwick, M.A. (eds) (2016) *Urban Landscape Ecology. Science, Policy and Practice.* Routledge.

53 Pittman, S.J., Kneib, R.T., and Simenstad, C.A. (2011) Seascape ecology: application of landscape ecology to the marine environment. *Marine Ecology Progress Series*, 427, 187–190.

54 Wedding, L.M., Lepczyk, C.A., Pittman, S.J., *et al.* (2011) Quantifying seascape structure: extending terrestrial spatial pattern metrics to the marine realm.*Marine Ecology Progress Series*, 427, 219–232.

55 Andrade, M. (2015) The influence of landscape and seascape features in the location of underwater recreation sites. Master´s Thesis in Landscape Architecture, Instituto Superior de Agronomia, Lisboa, Portugal.

56 Pittman, S.J. (2017) *Seascape Ecology: Taking Landscape Ecology into the Sea*, John Wiley & Sons, Ltd.

2

Points as Landscape Elements

2.1 The Different Patterns

The most important variable to characterize a group of elements represented by points in a landscape is its density (λ). The density of points in an area is simply measured as the number of points (n) divided by the total landscape area (TA) considered:

$\lambda = n/\text{TA}$

It is apparent, nevertheless, that for the same density of points, the ecological significance of these point elements in the landscape is also determined by their distribution or pattern. We can recall that the relationships between pattern and process are central to the discipline of Landscape Ecology.

However, before Landscape Ecology existed as a science, many plant ecologists were already concerned with the causes of patterns of distributions of plants at all scales, from those of individuals within a small area to those of vegetation types across the world. The importance of pattern was stressed by one of those first ecologists, Greig-Smith, who wrote in 1957 that "one of the principal contributions that may be expected from the use of quantitative methods in ecology is the more exact detection and description of distribution patterns"[1].

Animal ecologists have for a long time also attempted to establish relationships between distribution patterns and process (animal behavior). In many other disciplines the distribution pattern of points is of great interest. Geographers and geologists are often interested in the manner in which points (which may represent localities, oil wells, etc.) are distributed on a map. Statistical methods for data analysis of these point distributions are found in several books in the geology field, for example the classical work by Davis, first published in 1973[2].

In all fields where objects such as individual plants, animals, and oil wells can be located and represented on a map by points, the distribution of those points may be conveniently classified into three pattern categories: regular, random, and clustered (Figure 2.1). Of course, most points will have patterns intermediate between extreme regularity and extreme aggregation, and the question becomes one of determining where the observed pattern lies within the spectrum of possible distributions.

The detection and identification of the spatial patterns of plants and animals allows for the generation of hypotheses about the underlying causal factors or processes that may

Applied Landscape Ecology, First Edition. Francisco Castro Rego, Stephen C. Bunting, Eva Kristina Strand and Paulo Godinho-Ferreira.

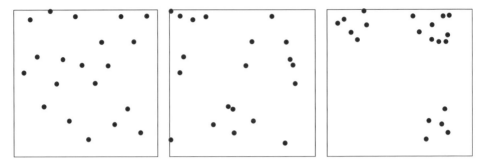

Figure 2.1 The three types of general patterns in the distribution of points, from regular (left) to random (center) and to clustered or aggregated (right).

Figure 2.2 Seed head of dandelion (*Taraxacum officinale*) with seeds easily dispersed by wind. *Source:* https://commons.wikimedia.org/wiki/File:Dandelion_seed_head_before_count.jpg; https://en. wikipedia.org/wiki/Seed_dispersal.

be responsible for those patterns. Some general considerations have been proposed for the causal processes explaining the patterns observed in ecological communities.

Random patterns generally imply environmental homogeneity and/or nonselective behavior of plants or animals. The absence of important causal factors or the random variation of underlying factors will both originate random patterns.

This generation of pattern can be illustrated by the process of seed dispersal. If seeds are easily dispersed by highly variable winds this factor might be approximated by a stochastic variation resulting in a random distribution of seeds (Figure 2.2). Also, in this case the location of each individual seed is not affected and does not affect the location of any other seed.

On the contrary, nonrandom patterns (regular or clustered) are indications that there are some important underlying factors with nonrandom spatial variation or simply that the location of an individual affects the location of other individuals.

A regular pattern can originate from a regularly spaced factor but it often results simply from negative interactions between individuals; that is, the location of one individual

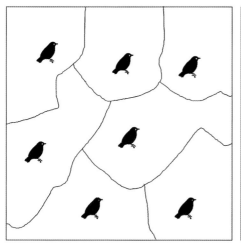 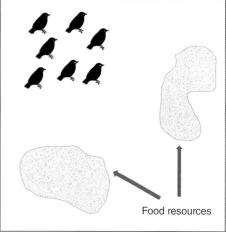

Food resources

Figure 2.3 Territorial behavior of birds is often associated with regularly distributed food resources while groups of birds are known to have advantages if food is distributed unpredictably and in clusters.

has a negative influence on the nearby placement of any other individual. This is particularly noticeable when the most important processes are the competition for limiting factors such as water, food, nutrients, light, or space, as expressed in the territorial behavior by animals (Figure 2.3) or the regular distance between neighboring plants.

A clustered pattern can be created when individuals aggregate in more favorable habitat patches. In plant communities it is generally observed that some of the most important underlying environmental factors (such as soil characteristics for plants or food resources for animals) are not randomly distributed in space. In this case the individual plants aggregate in the more favorable soil patches. Clustered patterns can also originate in gregarious behavior of animals or contagious processes in plants, whenever the location of an individual has a positive influence on the nearby location of other individuals. This can be simply illustrated by the clumping of new plants around an older plant, particularly noticeable when there is vegetative reproduction or when seeds are heavy and not easily dispersed (Figure 2.4).

We have demonstrated that simple factors such as seed dispersal may result in very different spatial patterns. However, even in this simple case of seed dispersal, it is not only the size of the seed but its propensity for wind dispersal that is important. In many cases seeds are dispersed by animals (Figure 2.5). This is particularly the case of frugivores, which move nonrandomly through the landscape guided by flesh-fruited trees and can generate spatially contagious seed dispersal in the preferred sites where they sleep, nest, lek, and perch[3]. Complex seed-disperser networks often result in spatially contagious seed dispersal[4].

A good example of these plant–animal interactions is provided by the random distribution of dwarf palms recolonizing old abandoned fields[5], explained by assistance of the long-distance dispersal carried out by the red fox, which deliver feces with seeds in a relatively scattered fashion.

Simple patterns can be easily identified and an association with hypothetical factors or processes can be tested. However, in many cases, several different factors may be

Figure 2.4 Cork oak (*Quercus suber*) with heavy acorns not easily dispersed by wind. *Source:* https:// upload.wikimedia.org/wikipedia/commons/6/66/Cork_oak,_Vale_da_ Azinheira,25_June_2016.jpg, https://commons.wikimedia.org/wiki/File:Quercus_ suber_g4.jpg.

Figure 2.5 Red foxes are known to disperse seeds of dwarf palms. Left: the Mediterranean dwarf palm (*Chamaerops humilis*). Photo by Tato Grasso (https://commons.wikimedia.org/w/index.php? curid=1333755). Right: red foxes (*Vulpes vulpes*) in England. https://upload.wikimedia.org/wikipedia/ commons/e/ee/Red_Fox_%28Vulpes_vulpes%29_-British_Wildlife_Centre-8.jpg.

influential and often they are not independent but exhibit complex interactions. However, since study of the causal factors determining the spatial pattern of the distribution of plants and animals is a prime objective of ecology, methods that can assist in detecting the patterns that developed through time are useful. The detection of pattern can be determined by two types of sampling methods, using distance methods (e.g. nearest neighbor) or quadrat analyses.

2.2   Distance Methods to Detect Pattern

In the simplest distance methods we perform a nearest-neighbor analysis where there is a comparison between the distances measured from points to their nearest neighbors and the distances that would be expected if the points were randomly placed.

In a large landscape (with minimum edge effects) the expected mean distance (EMD) between nearest neighbors under a completely random distribution is

$$EMD = \tfrac{1}{2}(TA/n)^{1/2}$$

where TA is the area of the map and n is the number of points. As the density of points (λ) is computed as the number of points (n) divided by the area of the map (TA) we can also compute EMD as

$$EMD = \tfrac{1}{2} (1/\lambda)^{1/2}$$

It is interesting to observe that, if the n points are regularly spaced in a squared grid in an area TA, the area of the square attributed to that point (the territory if it is an animal) is TA/n and its square root $(TA/n)^{1/2}$ is the side of the square or the distance between points. EMD is half of that value.

The observed mean nearest-neighbor distance (OMD) can be calculated as the average for all n points of the individual measured distances of each point (i) to its nearest neighbor (d_i):

$$OMD = \sum d_i/n$$

If the distribution is truly random, OMD and EMD should be the same. Therefore the ratio between EMD (the expected nearest-neighbor distance for a random distribution) and OMD (the observed distance) gives a nondimensional index of aggregation (PATTERN1) of the spatial pattern in the distribution of the points:

$$PATTERN1 = EMD/OMD$$

The value of PATTERN1 ranges from a minimum value of 0.465 for the maximum regularity (points arranged on a regular hexagonal pattern where every point is equidistant from six other points) to infinity (for a completely clustered distribution where all points coincide), going through the value of 1.0 for a completely random distribution of points (see Figure 2.6).

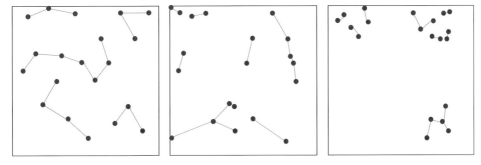

Figure 2.6 Nearest-neighbor distances for the example shown in Figure 2.1. For the same density of points, distances are larger in regularly spaced distributions (left), smaller for aggregated distributions (right), and intermediate in random distributions (center).

In summary:

> PATTERN1 < 1 regularity
>
> PATTERN1 = 1 randomness
>
> PATTERN1 > 1 clustering

For the example shown in Figure 2.1 with the distances presented in Figure 2.6 we can make some simple calculations. If there are 20 points in a landscape that represent 1 km × 1 km, then the density of points (λ) can be calculated as

$$\lambda = n/\mathrm{TA} = 20/1\mathrm{km}^2 = 20\,\mathrm{km}^{-2}.$$

The expected mean distance (EMD) under a random distribution would be

$$\mathrm{EMD} = \tfrac{1}{2}\left(1\mathrm{km}^2/20\right)^{1/2} = 0.112\mathrm{km}$$

The observed mean distances (OMDs) were calculated for the regular distribution at the left (OMD = 0.162 km), for the close to random distribution on the center (OMD = 0.116 km), and for the clustered distribution on the right (OMD = 0.074 km). The calculated indices were

> PATTERN1 = 0.69 (< 1 : regularity)
>
> PATTERN1 = 0.97 (close to 1 : randomness)
>
> PATTERN1 ≥ 1.51 > 1 : clustering

However, there is always the question of how close to unity should the value of PATTERN1 be for randomness to be assumed. To answer that question we can apply statistics to test the null hypothesis that the pattern is random[6]. The statistical test uses the difference between EMD and OMD and the normal distribution to compute z as

$$z = (\mathrm{EMD} - \mathrm{OMD})/\mathrm{SE}$$

where the standard error (SE) can be mathematically shown to be[7]

$$\mathrm{SE} = \left[(4-\pi)\,\mathrm{TA}/\left(4\pi n^2\right)\right]^{1/2} = 0.26136\left(\mathrm{TA}/n^2\right)^{1/2}$$

From this equation it becomes obvious that, for the same landscape area (TA), with an increasing number of points (n), the standard error decreases and it is easier to make statistical inferences about the true pattern.

It is therefore possible to conclude, with a probability of 95%, that if z is less than -1.96 this corresponds to observed distances much higher than expected than if the distribution was random, and the distribution is therefore likely to be truly regular. If the value of z lies between -1.96 and $+1.96$ we do not reject the hypothesis that the points are randomly distributed. If z is higher than 1.96 the observed distances are significantly lower than expected, suggesting a true clustered pattern.

In the example represented in Figure 2.6 with an area of TA = 1 km^2 and the number of points $n = 20$, the standard error can be computed as

$$\mathrm{SE} = 0.26136\left(1\mathrm{km}^2/20^2\right)^{1/2} = 0.013\,\mathrm{km}$$

The values of z for the three situations can now be estimated:

$z = -3.84 \, (< -1.96 \text{ significant regularity } 95\%)$

$z = -0.31 \, (\text{between } -1.96 \text{ and } +1.96 \text{ randomness not rejected})$

$z = 2.92 \, (> +1.96 \text{ significantly clustered})$

The statistical tests confirm the conclusions already presented. These analyses have been used in many fields to detect pattern and therefore infer process.

2.3 Quadrat Analysis to Detect Pattern

Another approach to detect pattern in the distribution of points is using quadrat analysis. Recall that, if we have n points in a total area TA, we can calculate the density λ as

$$\lambda = n/\text{TA}$$

If we divide the area into Q quadrats, each of area a, the total area TA is the product:

$$\text{TA} = a\,Q$$

The mean number of points per quadrat (MEAN) can be computed as the product of the density of points (λ) times the area of the quadrat (a). However, in practice, the mean is generally obtained by dividing the total number of points (n) by the total number of quadrats (Q):

$$\text{MEAN} = \lambda\,a = n/Q$$

We can now compute the variance in the number of points per quadrat as

$$\text{VARIANCE} = \sum (x_i - \text{MEAN})^2 / Q$$

where x_i is the number of points in the ith quadrat and Q is the number of quadrats. The pattern of the distribution can now be assessed by the VARIANCE/MEAN ratio (VMR) as another index of distribution:

$$\text{VMR} = \text{VARIANCE/MEAN}$$

The value of VMR can range from 0.0 for complete regularity (when all quadrats have the same number of points and variance is zero) to infinity (maximum clustering) going through 1.0, corresponding to complete randomness (according to a Poisson distribution):

$\text{VMR} < 1 \text{ regularity}$

$\text{VMR} = 1 \text{ randomness}$

$\text{VMR} > 1 \text{ clustering}$

An example of quadrat analysis using the distribution of points of the three landscapes presented in Figure 2.1 can illustrate the situation (Figure 2.7).

For all situations we have $n = 20$ points and an area of 1 km^2, and therefore the same density of points $\lambda = 20$ km^{-2} in the three landscapes. With the same number of quadrats $Q = 25$ we also have the same average number of points per quadrat

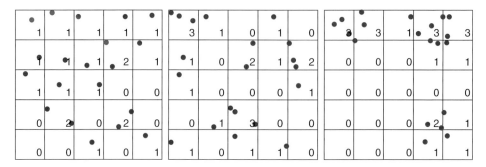

Figure 2.7 Counts of points per quadrat using the example of Figure 2.1 with three situations from regular to clustered distributions.

(MEAN) of $n/Q = 0.8$. The VARIANCE can be calculated for the three situations using the equation already presented $\left[\text{VARIANCE} = \sum(x_i - \text{MEAN})^2/Q\right]$. The three values obtained for the VARIANCE/MEAN ratio (VMR) were

$$\text{VMR} = 0.4/0.8 = 0.5 \,(<1\,\text{regularity})$$

$$\text{VMR} = 0.8/0.8 = 1.0 \,(=1\,\text{randomness})$$

$$\text{VMR} = 1.2/0.8 = 1.5 \,(>1\,\text{clustering})$$

It should be noted that both VMR and PATTERN1 have the value of 1.0 as the reference random distribution and in both indices lower values indicate regularity whereas higher values indicate clustering.

Similarly to what was done with nearest-neighbor analysis, we can also test in quadrat analysis the null hypothesis that the distribution was generated by a random process (VMR = 1.0).

We first have to calculate the standard error SE associated with VMR:

$$\text{SE} = \left[2/(Q-1)\right]^{\frac{1}{2}}$$

If Q is sufficiently large ($Q > 30$) and if we neglect edge effects, we can use the normal distribution and approximating z gives

$$z = (\text{VMR} - 1.0)/\text{SE}$$

Similarly to what was done before, it is now possible to determine, with a probability of 95%, that if the value of z is less than -1.96 this corresponds to smaller variances than expected in the random hypothesis and the distribution is likely to be regular. On the other hand, if z is higher than 1.96, the variance is higher than expected, suggesting evidence of a true clustered pattern. If the value of z lies between -1.96 and $+1.96$ we cannot reject the hypothesis that the points are randomly distributed.

In the above example the standard error SE was computed as

$$\text{SE} = \left[2/(Q-1)\right]^{\frac{1}{2}} = 0.29$$

We can now compute the values of z:

$$z = (0.5 - 1.0)/0.29 = -1.72 \text{ (non-significant tendency to regularity)}$$
$$z = (1.0 - 1.0)/0.29 = 0.00 \text{ (randomness assumed)}$$
$$z = (1.5 - 1.0)/0.29 = +1.72 \text{ (non-significant tendency to clustering)}$$

In this case, the z values did not show sufficient statistical evidence to allow a conclusion with high probability (>95%) that the distributions are not random. Nevertheless, the tendencies are obviously the same as detected by the other indices.

Some more detailed statistical analyses and test of hypotheses can be performed with these data. One of such analyses is in the comparison of the observed distribution of points with the expected distribution for a complete random pattern with the same mean value. This implies the use of a simple statistical distribution used for random processes, the Poisson distribution. We can now use the Poisson distribution to estimate the number of quadrats expected to have x points using the example of the small quadrat sizes.

The equation for the expected values of the number of quadrats n'_x having x points is

$$n'_x = Q\,\text{MEAN}^x \exp(-\text{MEAN})/x!$$

In the case of a number of $Q = 25$ quadrats and an average of $\text{MEAN} = 0.8$ points per quadrat we can compute the expected number of quadrats with x points (n'_x) if the distribution was truly random:

$$n'_0 = 25 \times 0.8^0 \times \exp(-0.8)/0! = 25 \times \exp(-0.8) = 11.2$$
$$n'_1 = 25 \times 0.8^1 \times \exp(-0.8)/1! = 25 \times 0.8 \times \exp(-0.8) = 9.0$$
$$n'_2 = 25 \times 0.8^2 \times \exp(-0.8)/2! = 25 \times 0.64 \times \exp(-0.8)/2 = 3.6$$
$$n'_{>2} = 25 - 11.2 - 9.0 - 3.6 = 1.2$$

The comparison of the expected values under the random (Poisson) distribution and the observed values is shown in Figure 2.8.

It can be easily concluded from the first graph that the observed number of empty quadrats (n_0) and quadrats with a higher number of points (n_x for x equal or higher than 2) is less than expected whereas the number of quadrats with one point (n_1) is higher than expected, indicating that the distribution shows a departure from randomness towards regularity. The second graph shows that the expected and observed frequencies are quite similar, indicating that the distribution is close to random. The third graph shows more extreme values with empty quadrats being more common than expected in a random distribution and quadrats with more than 2 points being more common than expected, indicating clustering.

A statistical test of the null hypothesis of randomness can be performed using a simple chi-square analysis comparing observed and expected counts for the different classes of number of points. The chi-square statistic χ^2 can be computed as

$$\chi^2 = \sum \left[(n_x - n'_x)^2 / n'_x \right]$$

In our example we have, for the three situations:

$$\chi^2 = (8-11.2)^2/11.2 + (14-9.0)^2/9.0 + (3-3.6)^2/3.6 + (2-1.2)^2/1.2 = 4.33$$
$$\chi^2 = (11-11.2)^2/11.2 + (10-9.0)^2/9.0 + (2-3.6)^2/3.6 + (2-1.2)^2/1.2 = 1.36$$

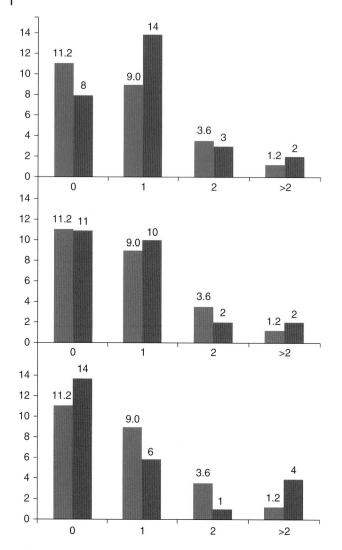

Figure 2.8 Comparison of the expected frequencies (Y-axis) for the number of quadrats with *Y* points under a random (Poisson) distribution with a total number of quadrats *Q* = 25 and a mean number of points per quadrat MEAN = 0.8 (in blue) with the observed frequencies (in red) for the three landscapes presented in the same order in Figure 2.1 and Figure 2.7.

$$\chi^2 = (14-11.2)^2/11.2 + (6-9.0)^2/9.0 + (1-3.6)^2/3.6 + (4-1.2)^2/1.2 = 10.11$$

These values can now be compared with the value of the chi-square distribution with a probability of 95% and 3 degrees of freedom (the number of classes minus one), which can be seen in any statistics book or table to be 7.82.

For the first landscape the chi-square value (4.33) is not higher than the corresponding value in the table (7.82) and therefore, in spite of the tendency for regularity, the null hypothesis of a random process cannot be rejected with this statistical test. For the landscape in the center the low chi-square value (1.36) indicates that there is almost no

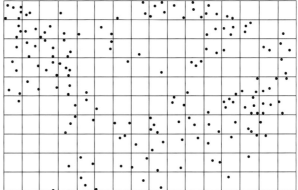

Figure 2.9 Pumpjack on an oil well in Texas (left). *Source:* Wikipedia, https://commons.wikimedia.org/w/index.php?curid=2351321. The geographical distribution of wells of the Permian Basin, Texas (right). Quadrats are 26 km^2 in size. *Source:* Davis, J.C. (1973) *Statistics and Data Analysis in Geology*, John Wiley & Sons, Inc., New York, NY.

departure from randomness. It is only in the third situation that the observed chi-square value (10.11) is higher than the threshold (7.82). In this situation this test allows for the conclusion that, with a high probability (95%), the distribution cannot be considered as random.

This type of analysis has been used in many fields to detect a pattern and then understand underlying processes. This is the case, for example, for the distribution of oil wells in Texas presented in one very influential work on statistics and data analysis in geology[8] (Figure 2.9).

In this example the area has been divided into 160 quadrats (Q) and has 168 wells (n), resulting in a value for the average number of points per quadrat of MEAN = 1.05. The VARIANCE was calculated to be 1.46. The VARIANCE/MEAN ratio (VMR) can then be computed as

$$VMR = 1.46/1.05 = 1.39$$

Since the variance is greater than the mean (VMR is greater than unity), we conclude that the pattern tends to be clustered. For the statistical test SE was calculated to be 0.112 and therefore

$$z = (1.39 - 1.00)/0.112 = 3.48$$

As this value (3.48) is higher than 1.96 we have some statistical evidence ($P > 0.95$) that the wells are truly clustered.

2.4 Consideration of Scale in Nearest-Neighbor Analyses

As in all landscape ecology studies, the scale of analysis is important. In detecting distribution patterns of points we used two approaches: nearest-neighbor distance and quadrat analysis. We will now indicate how scale can be included in distance (nearest-neighbor) analyses.

When using nearest-neighbor analysis only the closest point is considered. This has obvious limitations since a pattern is not fully defined by these first-order measurements.

If we want to measure spatial dependence of points based on their distances from one another we may want to use second-order point pattern statistics such as the *K*-function (Ripley's *K*), developed by Ripley in 1976[9] and widely used thereafter.

In this method circles of increasing radius (*r*) are placed around a point. The total number of neighbors within this radius is counted for each point and averaged over all points (or sample points) to give the observed number of points within radius *r* (ONR).

If the number of points in the circle follows a random distribution the expected number of points in a circle of radius *r* (ENR) would be a function of the overall density of points ($\lambda = n/\mathrm{TA}$) times the area of the circle (πr^2):

$$\mathrm{ENR} = \lambda \pi \mathrm{r}^2$$

We can transform this equation to get the radius of the circle that has ENR as the expected number of points:

$$\mathrm{r} = \left[\mathrm{ENR}/(\pi\lambda)\right]^{1/2}$$

We can now use a similar equation to calculate the radius of the equivalent circle *L*(*r*) with the same global density of points (λ) and having the observed number of points within radius *r* (ONR).

The value of *L*(*r*) is given by

$$\mathrm{L(r)} = \left[\mathrm{ONR}/(\pi\lambda)\right]^{1/2}$$

If the number of points observed within the radius *r* is higher than expected (ONR > ENR), then the circle of the same density having the same number of points will be bigger and *L*(*r*) > *r* and the difference *L*(*r*) − *r* is positive. This difference is generally used to understand pattern as a function of *r*:

$$\mathrm{f(r)} = \mathrm{L(r)} - \mathrm{r}$$

Positive values of this difference indicate that the radius of the equivalent circle is bigger than *r* and therefore points are clustered (aggregated) and negative values indicate that the equivalent circle is smaller than *r* as a result of a regular (dispersed) distribution (Figure 2.10).

In the example shown in Figure 2.9 the circles have radii that increase sequentially by 0.05 km. The observed (ONR) and the expected (ENR) number of points within each circle and the consequent calculations are presented in Table 2.1.

Typically the values of *f*(*r*) = *L*(*r*) − *r* are plotted against the values of *r* showing the patterns at the various scales (Figure 2.11).

These representations of pattern in the distribution of points have been used in many instances, as in the spatial distribution of trees. In this case it is common that regularity exists at shorter distances, indicating competition, but that aggregation occurs at larger distances as a function of site variation, as was detected in swamps with baldcypress (*Taxodium distichum*) (Figures 2.12 and 2.13).

A similar example can be found in the distribution of western juniper (*Juniperus occidentalis*) trees in the Owhyee Mountains, western USA (Figure 2.14).

In a study on the spatial distribution of the western juniper trees, Strand and others (2007) performed a Ripley´s *K* analysis, expressed in Figure 2.15.

Figure 2.10 Illustration of the method used to count neighbors within circles of increasing radius centered in three sample points applied to the regular distribution presented in Figure 2.1. If the points were randomly distributed, the number of points within the circles should be only a function of the overall density and the area of the circle.

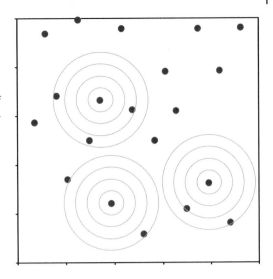

Table 2.1 Calculations of Ripley´s K analysis for the example shown in Figure 2.9. For the three sample points the circles of radius 0.05 km and 0.10 km have no neighbors, for the circles of radius 0.15 km we have a total of two neighbors (average 0.67 neighbors), and for the circles of radius 0.20 km we have a total of six neighbors (average 2.00 neighbors per sample point).

Radius (km) (r)	Expected number of neighbors (ENR)	Observed number of neighbors (ONR)	Radius of the equivalent circle $L(r)$	$L(r) - r$ $= f(r)$
0.05	0.16	0.00	0.00	−0.05
0.10	0.63	0.00	0.00	−0.10
0.15	1.41	0.67	0.10	−0.05
0.20	2.51	2.00	0.18	−0.02

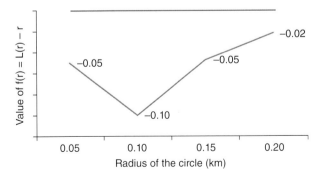

Figure 2.11 Results of Table 2.1 presented as a plot of $f(r) = L(r) - r$ against r showing negative values (regularity) at all scales, but especially at radius 0.10 km, indicating that within that distance there are much fewer neighbors than expected. The horizontal line (in red) indicates the value of zero associated with randomness.

Figure 2.12 Bald cypress (*Taxodium distichum*) trees in swamps of southeastern USA. *Source:* https://commons.wikimedia.org/wiki/File:Bald_Cypress.JPG.

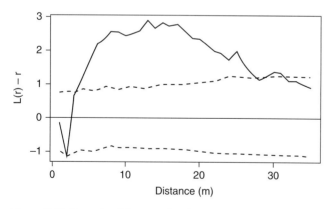

Figure 2.13 Example of the use of Ripley´s K analysis for the detection of pattern in the distribution of bald cypress (*Taxodium distichum*) trees in a swamp forest in South Carolina. The line with the value 0 represents randomness and the dashed lines represent the 95% upper and lower limits for the randomness hypothesis. The solid line represents the observed value of *L*(*r*) – *r* for the various distances. Negative values at shorter distances indicate regularity whereas positive values at larger distances mean clustering. The ecological processes causing these patterns are important questions for research. *Source:* Dixon, P.M. (2002) Ripley´s K function, in *Encyclopedia of Environmetrics* (eds A.H. El-Shaarawi and W.W. Piegorsch), John Wiley & Sons, Ltd, Chichester.

Figure 2.14 Juniper woodland expansion into sagebrush steppe in the Owhyee Mountains, USA. Photo by Stephen Bunting.

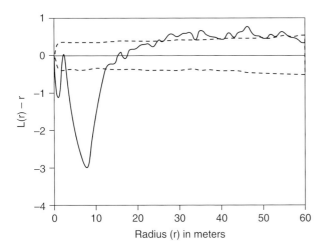

Figure 2.15 Plot of $L(r) - r$ versus r (distances in meters) using a spatial point pattern analysis and Ripley´s K to detect a pattern of western juniper trees[10]. At small distances negative values occur, indicating regularity, whereas at larger distances positive values show a trend to clustering.

From this analysis the authors[11] identified two different patterns at two statistically significant spatial scales that were a part of the encroachment process (Figure 2.15):

1) A regular pattern at short distances (less than 15 m) is probably associated with inhibition between juniper plants due to processes of competition for water and other resources. In fact, recruitment of young juniper occurred more often away from older plants, maximizing the utilization of water and light resources and perpetuating the spread of the species into previously juniper-free shrublands.
2) A significant clustering pattern within a 30–60 m radius of the juniper plants, attributed to the processes of seed dispersal by birds with small territories.

From this example we can conclude that the study of the process of seed dispersal and establishemt of new plants is important to understand and model observed patterns.

2.5 Consideration of Scale in Quadrat Analyses

An equivalent approach to include scale in a quadrat analysis has been used for a long time as it was known that the detection of randomness or aggregation depends on the size of the sample unit used[12]. In a very simple exercise Greig-Smith (1952)[13] illustrated this method by using a series of progressively larger quadrats to measure dispersion in an artificial situation with points representing the spatial distribution of plants (Figure 2.16).

In general, it is useful to plot the value of the VARIANCE to MEAN ratio (VMR) as a function of quadrat size. For another hypothetical example a plot is shown in Figure 2.16. If the pattern is detected at various scales, the processes that operate there could be better understood.

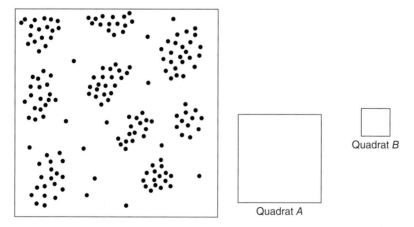

Figure 2.16 The measurement of pattern depends on the quadrat size used for the analysis. If quadrat A is used the analysis indicates that the population has a regular pattern, as aggregation occurs only at a smaller scale. However, if the analysis is done with quadrat B, the marked aggregation of the population at that scale would show. *Source:* Poole, R.W. (1974) *Introduction to Quantitative Ecology*, McGraw-Hill, New York, NY.

Figure 2.17 Photos showing the expansion of western juniper (*Juniperus occidentalis*) in eastern Oregon. The top photo is from 1890 and the lower photo was taken in 1989 (100 years of landscape change). *Source:* https://oregonstate.edu/dept/EOARC/pinon-juniper/material/Miller%20et%20al%202005.pdf.

We will again use the example of the expansion of western juniper in sagebrush steppe in the Great Basin USA (Figure 2.17) to perform a quadrat analysis and explore the relation between spatial patterns and ecological processes.

We can now perform the exercise using the variance/mean ratio presented earlier. The distribution of points (juniper trees) in an area of the same image of the aerial photograph of 1998 in Figure 2.18 can now be analysed by the variance/mean ratio, using quadrats of different sizes (Figure 2.19).

For the smaller (10 m × 10 m) quadrats we have the calculations illustrated in Table 2.2.

From the calculations in Table 2.2 we can compute:

$$\text{The MEAN} = \sum (n_x x) \Big/ \sum (n_x) = 437/800 = 0.55$$

$$\text{The VARIANCE} = \sum \left[n_x (x - \text{MEAN})^2 \right] \Big/ \sum (n_x) = 316.3/800 = 0.40$$

The counts for the 50 m × 50 m quadrats are presented in Table 2.3.

From the analysis of the counts of the 50 m × 50 m quadrats we computed the MEAN (13.66) and the population VARIANCE (21.04). Similarly, we computed means and variances for the 25 m × 25 m and 100 m × 100 m quadrats. The results are summarized in Table 2.4.

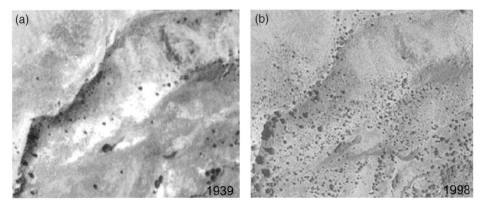

Figure 2.18 The same extraordinary expansion of western juniper (*Juniperus occidentalis*), as in Figure 2.17, can be observed in these aerial photographs from 1939 (a) and 1998 (b) taken in a 300 m × 400 m area in southern Idaho. Each dark gray dot on the photos is a juniper plant in the surrounding sagebrush steppe matrix (light gray) [14]. *Photo source*: www.earthexplorer.com; National Aerial Photography Program Data Directory.

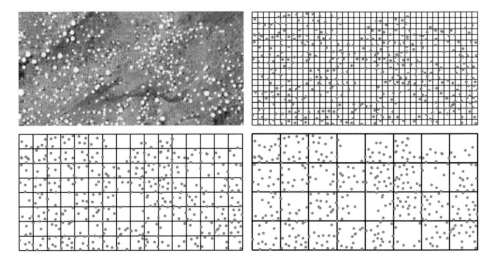

Figure 2.19 The aerial photograph (upper left) from the area shown in Figure 2.18(b) and different grids used for quadrat analyses. In this figure we show sides of quadrats of 10 m (top right), 25 m (bottom left), and 50 m (bottom right).

Table 2.2 Number of juniper trees per quadrat for the 10 m × 10 m quadrat in Figure 2.18.

Number of trees per quadrat (x)	Number of quadrats (n_x)	Number of trees ($n_x x$)	$n_x(x - \text{MEAN})^2$
0	418	0	124.7
1	331	331	68.1
2	47	94	99.3
3	4	12	24.1
	$\sum(n_x) = 800$ (total number of quadrats)	$\sum(n_x x) = 437$ (total number of trees)	$\sum[n_x(x - \text{MEAN})^2] = 316.3$

Table 2.3 Number of trees in each of the 50 m × 50 m quadrats presented in the bottom right of Figure 2.19.

18	15	14	4	12	15	9	5
11	16	9	16	21	20	7	16
13	11	19	8	10	19	16	16
14	21	14	14	18	6	12	18

Table 2.4 Example of changes in the variance/mean (VMR) ratio for different quadrat sizes.

Quadrat size	MEAN	VARIANCE	VMR
10 m × 10 m	0.55	0.40	0.72
25 m × 25 m	3.42	3.08	0.90
50 m × 50 m	13.66	21.04	1.54
100 m × 100 m	54.63	91.23	1.67

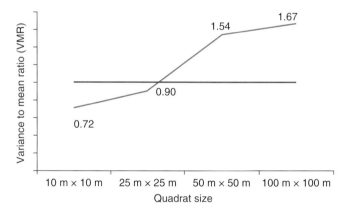

Figure 2.20 VARIANCE to MEAN ratio (VMR) for different quadrat sizes used in the example shown in Table 2.4. The horizontal line (in red) represents complete randomness (VMR = 1.0). Values of VMR above and below unity for smaller quadrat sizes indicate regularity, whereas values above unity for larger quadrat sizes show aggregation.

Obviously the values of the means are different as the sizes of the quadrats are different. The mean for the larger quadrat size is 100 times that of the smaller quadrats, and the variance also increases from smaller to larger quadrats. However, the ratio variance to mean can show different values at different quadrat sizes, depending of the pattern of the distribution, as shown in the example in Table 2.4 and Figure 2.20.

The conclusions of the quadrat analysis performed in a small area of the juniper distribution in eastern Oregon confirm the patterns that were already detected using

Ripley's *K*, that is, a shift from a regular pattern at short distances (VMR < 1 at 10 m) to a clustering pattern at higher distances (VMR > 1 at 50 m), going through a pattern close to random at intermediate distances (VMR closer to 1 at 25 m). The ecological interpretations of these findings are the same as before.

The examples presented show how patterns can be measured by distance and quadrat methods, how indices are dependent on scale, and how this detection of pattern can elucidate the processes affecting point elements in the landscape.

Key Points

- "One of the principal contributions that may be expected from the use of quantitative methods in ecology is the more exact detection and description of distribution patterns" (Greig-Smith, 1957)[15].
- Distribution patterns and ecological processes that create them are related to each other.
- The distribution of points may be conveniently classified into three pattern categories: regular, random, and clustered.
- A regular pattern is created by processes where the placement of a point has a negative influence on the nearby placement of any other point, a random pattern can be created if the placement of any point has no influence on the placement of any other point, and a clustered pattern is created by processes where the placement of a point has a positive influence on the nearby placement of any other point.
- The detection of pattern can be determined by two sampling methods. The nearest-neighbor (distance) method is a comparison between the distances measured between points and their nearest neighbors with the distances that would be expected if the points were randomly placed. The quadrat method assessed the variance/mean ratio of the number of points per quadrat as an index of clustering.
- In both approaches the issue of scale is important. The detection of pattern depends on the size of the sample unit used (the number of order of the considered distance between points or the size of the quadrat).
- Depending on scale we obtain a distribution pattern instead of a single value.

Endnotes

1 Greig-Smith, P. (1957) *Quantitative Plant Ecology*, Academic Press, New York and Butterworths Scientific Publications, London.
2 Davis, J.C. (1973) *Statistics and Data Analysis in Geology*, John Wiley & Sons, Inc., New York, NY.
3 Beckman, N.G. and Rogers, H.S. (2013) Consequences of seed dispersal for plant recruitment in tropical forests: interactions within the seedscape. *Biotropica*, 45, 666–681.
4 Fedriani, J.M. and Wiegand, T. (2014) Hierarchical mechanisms of spatially contagious seed dispersal in complex seed-disperser networks. *Ecology*, 95, 514–526.

5 Jácome-Flores, M.E., Delibes, M., Wiegand, T., and Fedriani, J.M. (2016) Spatial patterns of an endemic Mediterranean palm recolonizing old fields. *Ecology and Evolution*, 6, 8556–8568.

6 Krebs, C.J. (1989) *Ecological Methodology*, Harper and Row, New York, NY.

7 Clark, P.J. and Evans, F.C. (1954) Distance to nearest neighbor as a measure of spatial relationships in populations. *Ecology*, 35, 445–453.

8 Davis, J.C. (1973) *Statistics and Data Analysis in Geology*, John Wiley & Sons, Inc., New York, NY.

9 Ripley, B.D. (1976) The second-order analysis of stationary point processes. *Journal of Applied Probability*, 13, 255–266.

10 Strand, E.K., Robinson, A.P., and Bunting, S.C. (2007) Spatial patterns on the sagebrush steppe/Western juniper ecotone. *Plant Ecology*, 190, 159–173.

11 Strand, E.K., Robinson, A.P., and Bunting, S.C. (2007) Spatial patterns on the sagebrush steppe/Western juniper ecotone. *Plant Ecology*, 190, 159–173.

12 Poole, R.W. (1974) *Introduction to Quantitative Ecology*, McGraw-Hill, New York, NY.

13 Greig-Smith, P. (1952) The use of random and contiguous quadrats in the study of the structure of plant communities. *Annals of Botany*, 16, 293–316.

14 Strand, E.K., Robinson, A.P., and Bunting, S.C. (2007) Spatial patterns on the sagebrush steppe/Western juniper ecotone. *Plant Ecology*, 190, 159–173.

15 Greig-Smith, P. (1957) *Quantitative Plant Ecology*, Academic Press, New York, and Butterworths Scientific Publications, London.

3

Linear Elements and Networks

3.1   The Linear Features and Corridors in the Landscape

Linear features in the landscape are commonly created both by nature, such as streams, ridges, and animal trails, or by humans, such as roads, powerlines, ditches, and walking trails[1]. These linear features affect many ecological characteristics and processes in a landscape, including the influence of wind, solar radiation, the movement of disturbances, or the movement of organisms.

One can consider two categories of linear features: line corridors and strip corridors. Both types of linear features may have the functions of conduits (corridors), barriers or filters for the movement of organisms, disturbances, or for the flow of water, soil, or nutrients. However, strip corridors differ from line corridors in that their width is important for the process or organism of interest, as when they are sufficiently wide to provide habitat and also functions in the landscape, acting as a source or sink for organisms, matter, or disturbances.

Since the difference between line corridors and strip corridors is whether or not they provide habitat (and therefore possible source or sink functions), the distinction is often only a matter of scale and the objective of the analysis.

Literature about corridors is abundant and rapidly increasing. It typically concentrates on riparian vegetation, hedgerows, and roads, but studies have also considered how railroads, dikes, ditches, fences, powerlines, and vegetation strips influence wildlife movements.

There are many references on such studies from the very beginning of landscape ecology. The study on the ecological effects of roads have been developing in many countries[2], especially after Forman published the very influential book reviewing our knowledge on the effects of roads on organisms and processes at the landscape level in what was referred to as "road ecology"[3].

Many of the studies on corridors are dedicated to their function as conduits for organisms, for the dispersal of plant seeds, and for the movement of animals by facilitating the movement through otherwise unsuitable habitat. They may also act as transportation or recreation pathways for humans (Figure 3.1).

In contrast, corridors may also serve as barriers to organism movements, including humans, depending on the characteristics of the corridor and the requirements of the organism in question. The filtering function of corridors can be likened to acting as a leaky barrier that restricts the movements of some organisms but not others. This

Applied Landscape Ecology, First Edition. Francisco Castro Rego, Stephen C. Bunting,
Eva Kristina Strand and Paulo Godinho-Ferreira.
© 2019 John Wiley & Sons Ltd. Published 2019 by John Wiley & Sons Ltd.

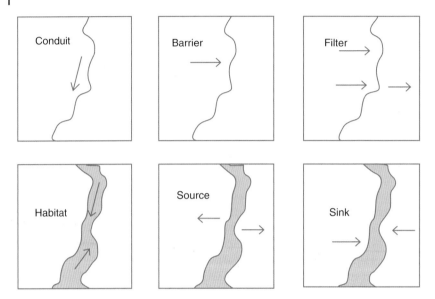

Figure 3.1 Ecological functions of line (and strip) corridors, which may act as conduits, barriers, or filters (top row). In addition, strip corridors (bottom) may also act as habitat, sources, or sinks (adapted from Smith, D.S. and Hellmund, P.C. (1993) *Ecology of Greenways: Design and Function of Linear Conservation Areas*, University of Minnesota Press, Minneapolis, MN).

filtering function is not limited to organisms. In riparian zones along stream corridors, eroded soil can be trapped and chemicals can be filtered out of soil water by plant roots and soil microbes. They may also act as a filter to disturbances. For example, fire may cross a strip under some types of weather conditions but not others.

Greenways are special types of corridors that are implemented in or near urban environments and function primarily as a conduit for humans with a focus on recreation and aesthetics[4], but that may also provide many other ecological functions (Figure 3.2).

Various types of greenways were already described in 1990 by Charles Little in his fundamental book *Greenways for America*: "a linear open space established along either a natural corridor, such as a riverfront, stream valley, or ridgeline, or overland along a railroad right-of-way converted to recreational use, a canal, scenic road or other route; a natural or landscaped course for pedestrian or bicycle passage; an open-space connector linking parks, nature reserves, cultural features, or historic sites with each other and with populated areas; locally certain strip or linear parks designated as parkway or greenbelt"[5].

Possibly the first clear example of a greenway was that created in 1880 by Frederick Law Olmsted in his "Emerald Necklace" plan for the Boston Park System (Figure 3.3).

The modern study of the ecology of greenways[6] began in 1993 but, since the very beginning, the word "greenways" has been used with many different meanings. However, in spite of some confusion around its definition, the concept gained popularity and appeared regularly in popular language and planning policy in the USA. An international movement was then beginning, as announced by Julius Fabos and

Figure 3.2 Greenway in Moscow, Idaho, USA. Photo by Stephen Bunting. http://mapio.net/o/865258

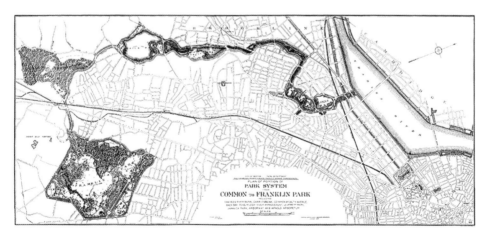

Figure 3.3 The Olmsted Plan "Emerald Necklace", a greenway in the urban area of Boston. *Source:* https://en.wikipedia.org/wiki/Emerald_Necklace.

Jack Ahern in 1996[7]. A good review of the use of greenways for strategic landscape planning was provided by Ahern in 2002[8].

The linear features in the landscape, whatever their function and context will be, need to be characterized in order to make it possible to understand their relationships with landscape processes.

3.2 Curvilinearity and Fractal Analysis

We will start the analysis of linear elements by the simple case of one single line (no width). This seems simple in that the only characteristic that is generally quantified is the length. Of course, length is a very relevant measurement with very important ecological consequences. The length of a river, for instance, determines the amount of adjacent riparian vegetation. The length of a road determines the number of houses that can be built adjacent to it if the distance between them is fixed. The length of a coastal line also determines the amount of habitat coastal nesting birds have to utilize.

However, the measurement of the length is not as simple as it seems. The "length" of a coastline was considered to be a very difficult characteristic and not a meaningful measure in many studies, such as that of Pennycuick and Kline[9] when they studied the distribution of bald eagle nests in the cliffs of the coasts of the Aleutian Islands of Alaska (Figures 3.4 and 3.5).

In fact, most attempts to measure density of organisms that are distributed along irregular boundaries, such as coastlines, have the fundamental difficulties associated with the unit of measurement. Therefore we need, again, to look at the scale appropriate to the organism or the process of interest.

A common example illustrating the importance of scale and the measurement unit is the estimation of the length of the coast of Britain (Figure 3.6). Here, again, the coastline

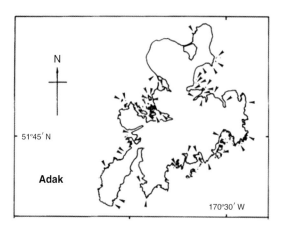

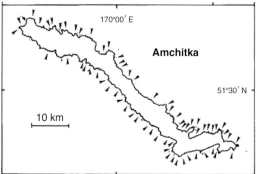

Figure 3.4 Maps of Adak and Amchitka Islands in Alaska, USA. The arrows point to bald eagle (*Haliaeetus leucocephalus*) nest sites. *Source:* Pennycuick, C.J. and Kline, N. C. (1986) Units of measurement for fractal extent, applied to the coastal distribution of bald eagle nests in the Aleutian Islands, Alaska. *Oecologia*, 68, 254–258.

Figure 3.5 Bald eagle (*Haliaeetus leucocephalus*) in its nest in the coastal habitat. *Source:* http://
www.arkive.org/bald-eagle/haliaeetus-leucocephalus/image-G54270.html; http://cdn1.arkive.org/
media/67/678A6C5F-678B-4B13-8711-3171264AF87E/Presentation.Large/Bald-eagle-in-nest-in-
coastal-habitat-.jpg.

Figure 3.6 Land's End, at the tip of Cornwall, Great Britain. *Source:* https://en.wikipedia.org/wiki/Land%
27s_End.

Figure 3.7 Benoit Mandelbrot (1924–2010), the pioneer of fractal geometry. *Source:* https://upload. wikimedia.org/wikipedia/commons/thumb/e/e3/Benoit_Mandelbrot,_TED_2010.jpg/220px- Benoit_Mandelbrot,_TED_2010.jpg.

is difficult to define. In order to answer the question we have to realize that the response depends on how closely you look at it or how long your measuring ruler is. The answer is that the length of the coastline seems to get longer and longer as you measure it more closely.

We can now use the concept of fractal dimension to describe the coastline, as did Mandelbrot (Figure 3.7) in 1967 is his very influential article on fractals on "how long is the coast of Britain?"[10].

In order to see how the measured length changes with scale we need to measure the coastline using rulers as segments of different lengths, as if we were considering different species of territorial birds that would require different distances between nests.

Let us define the length of the ruler as a segment of size s and that we measure the coastline as the number of segments (n) (Figure 3.8).

With a larger ruler (the first image to the left) the segment size is $s = 400$ km, the measured number of segments is $n = 9$, and the estimated length is therefore $ns = 3600$ km. For the 200 km ruler the number of segments is 19 and $ns = 3800$ km, for the 100 km ruler $n = 48$ and $ns = 4800$ km, and for the smaller ruler ($s = 50$ km) the measured number of segments is $n = 97$ and the estimated length of the coastline is $ns = 4850$ km.

With a perfectly linear coastline the estimated lengths should be similar (L) and therefore $ns = L$ or $n = L/s$. For the more general case, however, we have

$$n = L/s^D = Ls^{-D}$$

where L is a constant (the length of a reference line), s is the size of the ruler, and D is termed the fractal dimension of the line, which is a measure of its curvilinearity. If the line is not curvilinear (a straight line) the value of D will be unity and we will have the simple

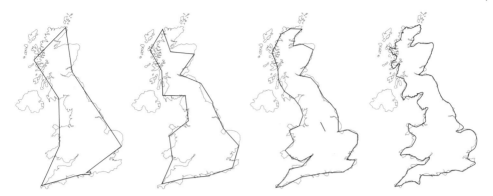

Figure 3.8 Measurements of the coastline of Great Britain using different ruler lengths, varying from 400 km, 200 km, 100 km, and 50 km, from left to right.

case where $n = L/s$. The closer to one the dimension D is, the straighter and smoother the coastline is. The higher the dimension (the closer to two) is, the more jagged and wiggly the coastline is[11].

If we take the natural logarithm of the measured values we can fit a straight line of the form:

$$\ln(n) = \ln(L) - D \ln(s)$$

In this case the fractal dimension of the coastline, D, is simply the slope of the line.

It is easier to see this relationship in a graph where the number of segments needed to measure the coast of Britain (shown in Figure 3.8) is plotted against the size of the ruler used (Figure 3.9).

We can also fit a power equation to the relation between n and s and find that the exponent is approximately $D = 1.163$, which is our estimate of the fractal dimension of the coastline of Britain.

The final equation for the length of the coastline measured in number of segments is

$$n = (L)\left(s^{-D}\right) = 9454\,s^{-1.163}$$

and the estimated length ns is therefore

$$ns = (L)\left(s^{1-D}\right) = 9454\,s^{-0.163}$$

The value of k (9454 km) is the estimate of the length of the coast measured with a 1 km ruler. As the exponent of s is negative, this equation shows that the estimated length of the coastline decreases with the increasing size of the ruler (s). If the coastline was a straight line the value of D would be 1 and the measured length would always be L independently of the size of the ruler.

We can now use this equation to estimate how many objects there could be along the coastline knowing the required distance from each other (s). Scale is now included in the solution. The length of the coastline depends upon the ruler used to measure it and the ruler depends on the organism of interest (distance between bald eagle nests, for example).

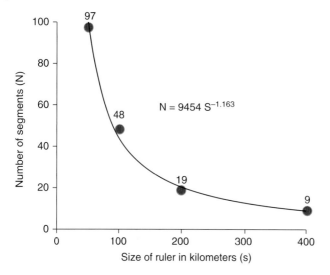

Figure 3.9 Relationship between the number of segments needed for the measurement of the coast of Britain and the size of the ruler used (in kilometers). The power equation fitted is shown, indicating that the fractal dimension (D) is equal to 1.163.

Similarly to what was done for points in the previous chapter, we can apply methods that use distances (as rulers in this case) or quadrats. In this latter case we can use the box counting method, which is analogous to the length measuring method previously used (Figure 3.10). We start by using a large grid (400 km side) and count the number of boxes (quadrats) of the grid where the coastline is included. Then we repeat the measurement using successively smaller boxes.

Similarly to what was done with the ruler, we can now plot n against s, where s is now the side of the box, as shown in Figure 3.11.

The results of both methods indicate similar values for the slope ($D = 1.163$ and $D = 1.178$, respectively). The distance (ruler) and quadrat (box counting) methods are equivalent in determining the fractal dimension of lines, that is, their curvilinearity.

Rivers, when reduced to lines, indicate the same effects of measurement scale as coastlines (Figure 3.12). The shape of the river has important consequences on hydraulic and biotic processes. Greater curvilinearity usually results in a decreased stream gradient, which influences the erosion power and potential sediment load capacity of a river. Sinuous streams have a greater riparian habitat and often more diversity in the types of habitat. Curvilinearity may also influence stream characteristics such as temperature, oxygen content, and productivity.

The concept of a fractal dimension can be used to describe the complexity of all shapes. From this example we understand that as a line changes from straight to a more irregular shape the fractal dimension D changes from 1 and approaches the value of 2 as the line fills the whole area. Using the same concept we can include a third dimension and as the irregularity of the surface increases the fractal dimension D increases from 2 and approaches the value of 3 as the surface fills the whole volume. Likewise, as a straight line is fragmented, the fractal dimension D approaches 0. The value of 0 is the fractal dimension of a point. This is illustrated in Figure 3.13.

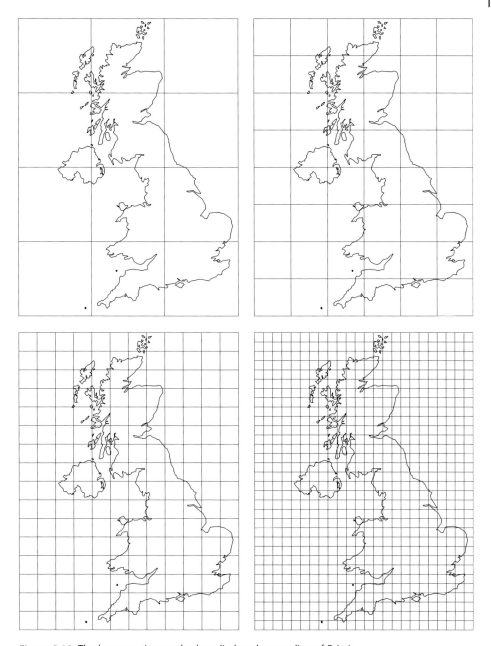

Figure 3.10 The box counting method applied to the coastline of Britain.

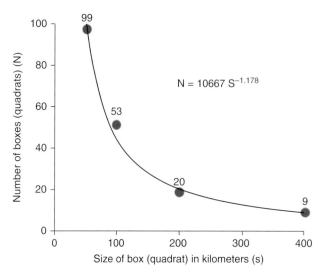

Figure 3.11 The best fit for a power equation for the results of the box counting method for the coastline of Britain. The estimate of the fractal dimension of the coastline of Great Britain is therefore $D = 1.178$.

Figure 3.12 Image of strong curvilinearity of a river close to Faro, Portugal. *Source:* Image courtesy of REDDIT user Docious.

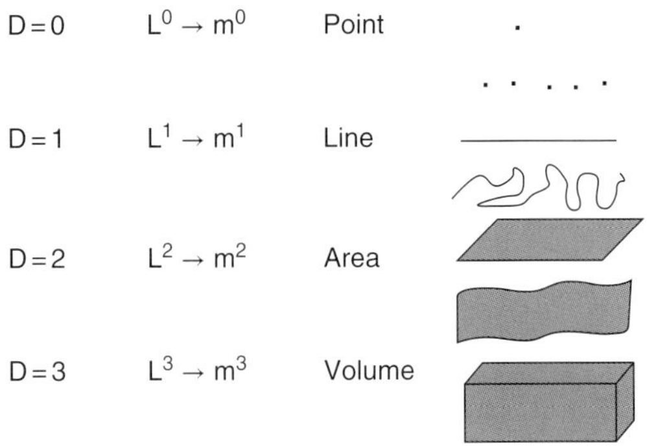

Figure 3.13 Representation of shapes from the more simple (a point with fractal dimension $D = 0$, nondimensional) to the more complete (a volume with fractal dimension $D = 3$, with units of length L^3, commonly measured in cubic meters, m^3).

$D = 0$	$L^0 \rightarrow m^0$	Point	
$D = 1$	$L^1 \rightarrow m^1$	Line	
$D = 2$	$L^2 \rightarrow m^2$	Area	
$D = 3$	$L^3 \rightarrow m^3$	Volume	

3.3 Linear Density of Networks

We will now consider a network of lines. Of course, individual lines can be quantified by the method presented earlier. However, there are several attributes that can be used specifically to describe a network of linear elements in a landscape.

The most important single characteristic is the line (or corridor) density, as a measure of the abundance of corridors in an area, which can be simply defined as the total length of linear elements per unit area. The line density (λ_l) is given by

$$\lambda_l = \sum l_i/\text{TA} = \text{TL}/\text{TA}$$

where l_i is the length of line i and TL is the total length of lines within the area TA.

For watersheds this concept is used in defining drainage density as the total length of streams divided by the area of the drainage basin. Drainage density is often reported in units of km/km^2 or km^{-1} and values from less than 2 km^{-1} to over 100 km^{-1} have been reported as a function of climate, soils (less permeable soils result in more surface water runoff and higher drainage density values), and topography (watersheds with steep slopes tend to have a higher drainage density).

It can be shown that the average distance from the basin divide to a stream is $1/(2\lambda_l)$ and that the average distance that a drop of water from a rainfall event travels to a stream is also inversely proportional to the drainage density. Watersheds with a high drainage density have a shorter response time to a precipitation event and a sharper peak discharge. Thus, drainage density is an indicator of the efficiency of the stream network in the process of draining an area and it has been used in the predictions of the magnitudes of flood flows or low flows (Figure 3.14)[12].

Drainage density applies for the length of the streams, but the concept of linear density can also apply to any other network of lines on a map. We can also define the linear density of ridges and of contour lines. In watershed studies the density of contour lines is directly related to the average slope of the watershed. However, the density of contour lines has also been used to estimate landscape ruggedness relevant for many different processes, such as for the "escape terrain" for bighorn sheep (*Orvis canadensis*)[13,14] (Figure 3.15).

The effects of linear networks in ecological processes are many and are dependent on the type of network and the process considered. One such cause–effect relationship that

Figure 3.14 Areas with different drainage densities, increasing from left to right. *Source:* fao.org (Food and Agricultural Organization of the United Nations).

Figure 3.15 Bighorn sheep (*Ovis canadensis*) in rugged terrain. *Source:* https://wallpaper.wiki/colorado-mountains-background-hd.html/wallpaper-wiki-colorado-mountains-desktop-background-pic-wpc005599.

has been abundantly reported in the literature is the effect of road density, a linear network, in inhibiting movement of large wildlife species. These results have been used to close some roads and to propose specific forest management schemes in order to mitigate the impact of road networks on sensitive wildlife species[15].

Some wildlife species have been shown to be particularly affected by road density. Sensitivity to road density has been demonstrated in a study on grizzly bears (*Ursus arctos horribilis*) in Idaho, concluding that open roads greatly influenced the distribution of bears (Figure 3.16). Areas with high open road density (>2.6 km^{-1}) were avoided or used much less than expected if the landscape had been randomly used[16].

Similar studies were conducted in other regions. Recent research has shown the negative impact of road density on survival rate of grizzly bear populations in Alberta, Canada (Figure 3.17)[18].

Figure 3.16 Grizzly bear (*Ursus arctos horribilis*) image from the report of the study in Idaho[17]. *Source:* http://igbconline.org/wp-content/uploads/2016/02/Wakkinen_Kasworm_1997_Grizzly_bear_and_road_density_relation.pdf.

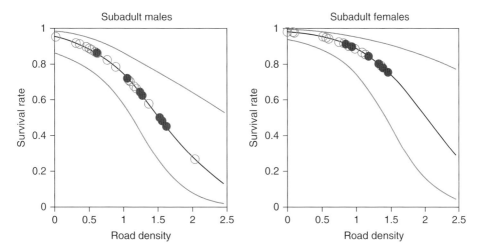

Figure 3.17 The negative relationship between road density (km^{-1}) and survival rate of subadult males and females in a study of grizzly bear in Alberta, Canada. *Source:* Boulanger J. and Stenhouse, G.B. (2014) The impact of roads on the demography of grizzly bears in Alberta. Open-access article. *PLoS ONE*, 9 (12), e115535.

3.4   Spatial Distribution of Linear Networks

Apart from linear density, linear networks can also be described by their spatial distribution, as they are also known to affect landscape processes on many scales. We could use quadrat methods in exactly the same way that they were used for points. In that case a fractal dimension for the network of lines can be determined exactly as if it was a single line. However, the quantification of the spatial distribution of lines is less well developed and used than those applied to the pattern of points. One of the most common methods to examine the distribution of distances between lines is by calculating nearest-neighbor distances[19].

The steps required for this technique include:

1) Randomly pick a point on one of the lines (i) in the landscape.
2) From that point in line (i), measure the distance to the nearest line in a direction perpendicular to the line (d_i), as shown in Figure 3.18.
3) Randomly pick other points on other lines and repeat the procedure.
4) Compute the observed mean nearest-neighbor distance (OMD) for the n distances measured:

$$\text{OMD} = \sum d_i/n$$

5) Compute the expected mean nearest-neighbor distance (EMD) for a pattern of random lines with the equation[20]

$$\text{EMD} = 1/(\pi\lambda_1) = 0.31831/(\lambda_1)$$

where λ_1 is the line density.

6) Compute a nearest-neighbor index (PATTERN2) identical to that used for point patterns by taking the ratio

$$\text{PATTERN2} = \text{EMD/OMD}$$

Similarly to what was indicated for point distribution, the meaning of the index is

PATTERN2 < 1 regularity
PATTERN2 = 1 randomness
PATTERN2 > 1 clustering

The calculations are illustrated in Figure 3.19.

If the total length of these lines (TL) is 43 km and the area (TA) is 100 km^2 the linear density is calculated as

$$\lambda_1 = \text{TL/TA} = 43\,\text{km}/100\,\text{km}^2 = 0.43\,\text{km}^{-1}$$

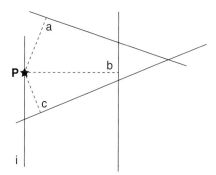

Figure 3.18 The distance of the random point P in line i to the nearest line is $d_i = c$.

Figure 3.19 Distribution of lines used for the exercise, where l_i is the length of line i and d_i is the nearest-neighbor distance from a point on line i.

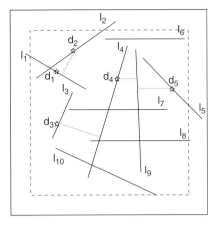

If the $n = 5$ distances measured (km) inside the inner area (to avoid edge effects) were $d_1 = 0.2$, $d_2 = 1.2$, $d_3 = 2.1$, $d_4 = 1.0$, and $d_5 = 1.8$, we would compute the observed mean nearest-neighbor distance (OMD) as

$$\text{OMD} = \sum d_i/n = 6.3\,\text{km}/5 = 1.26\,\text{km}$$

We can now compute the expected mean nearest-neighbor distance (EMD in the hypothesis of a random distribution) as

$$\text{EMD} = 0.31831/0.43\text{km}^{-1} = 0.74\,\text{km}$$

and the ratio PATTERN2 can now be calculated as

$$\text{PATTERN2} = \text{EMD}/\text{OMD} = 1.26\,\text{km}/0.74\,\text{km} = 1.70$$

As the value of PATTERN2 is well above unity (PATTERN2 > 1) we can conclude that the distribution of lines tends to be clustered.

An alternative system to detect patterns in the distribution of lines is by constructing a number of auxiliary transect lines over the region of interest, counting the number of times these auxiliary lines cross the lines of the network, and measuring the distance between those lines (d_i) (Figure 3.20).

Figure 3.20 Distribution of lines used for the exercise, where l_i is the length of line i and d_i is the distance between lines.

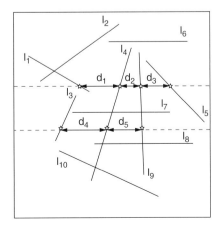

The same example would indicate that the measured distances (km) are now $d_1, d_2, d_3,$ $d_4,$ and d_5. The variance of these distances is also a good measure of departure from absolute regularity when the variance is zero.

The relationships between the total number of crossings (CROSS) and the total length of the auxiliary lines (TLA) is also very important since it allows for the estimation of linear density by $\lambda_l = \pi$ CROSS/(2 TLA). In this case, with 5 crossings (CROSS = 5) and a total length of the two auxiliary lines TLA = 20 km, we would estimate the linear density to be 0.39 km^{-1}, close to the measured value of 0.43 km^{-1}. These relationships were established and used successfully for stream networks[21]. These methods work well for lines that do not frequently reverse directions and that are at least 1.5 times longer than the average distance between lines[22].

This method of using auxiliary transect lines is extremely useful as it can provide information simultaneously about linear density and distribution.

3.5 Analysis of the Spatial Distribution of Linear Networks

To illustrate the process fully we can use a worked example from the northern coast of the island of Terceira, Azores, where the traditional agricultural fields are separated by stone walls (from lava flows) or hedgerows with woody species such as the firetree (*Myrica faya*) (Figure 3.21).

We can now analyse the pattern of the lines that separate the agricultural fields using the images of the same general region from Google Earth (Figure 3.22).

In a way similar to what was done for point counts in quadrats, we can now count the number of crossings between the segments in the transect lines and the edges of the

Figure 3.21 A view of the area of Biscoitos, on the island of Terceira, Azores, showing the stone walls and the hedgerows separating the small agricultural fields. *Source:* Manuel Cunha, SKai aerial image, http://i.ytimg.com/vi/BLh2d8qFzYM/maxresdefault.jpg.

Figure 3.22 A view of the northern coast of the island of Terceira, Azores, close to Biscoitos. We used four auxiliary transect lines divided into 112 segments of 50 m (upper image) and 56 segments of 100 m (lower image). *Source:* Google Earth.

fields. A summary of the results obtained and the calculations of the mean and variance are presented in Tables 3.1, 3.2, and 3.3.

As the variance to mean ratio (VMR) is well below unity, we conclude that the lines (the limits of the agricultural fields) were regularly distributed at both scales (100 m and 50 m).

As for points, we could now calculate the expected number of segments under the null hypothesis of a random (Poisson) distribution with the same mean values, and test that hypothesis with a chi-square test comparing observed and expected counts.

Table 3.1 A summary of the crossings of the transects with 100 m segments.

Number of crossings per segment (x)	Number of segments (n_x)	(n_x)(x)	(n_x)(x − MEAN)²
0	1	0	3.5
1	17	17	12.6
2	28	56	0.5
3	9	27	11.7
4	1	4	4.6

With:

$$\sum (n_x) = 56 \text{ (total number of segments)}$$

$$\text{CROSS} = \sum [(n_x)(x)] = 104$$

$$\text{MEAN} = \sum [(n_x)(x)] / \sum (n_x) = 104/56 = 1.86$$

$$\sum [(n_x)(x - \text{MEAN})^2] = 32.9$$

$$\text{VARIANCE(of the population)} = \sum [(n_x)(x - \text{MEAN})^2] / \sum (n_x) = 32.9/56 = 0.59$$

Table 3.2 A summary of the crossings of the transects with 50 m segments.

Number of crossings per segment (x)	Number of segments (n_x)	(n_x)(x)	(n_x)(x − MEAN)²
0	25	0	21.6
1	71	71	0.3
2	15	30	17.2
3	1	3	4.3

With:

$$\sum (n_x) = 112 \text{(total number of segments)}$$

$$\text{CROSS} = \sum [(n_x)(x)] = 104$$

$$\text{MEAN} = \sum [(n_x)(x)] / \sum (n_x) = 104/112 = 0.93$$

$$\sum [(n_x)(x - \text{MEAN})^2] = 43.4$$

$$\text{VARIANCE(of the population)} = \sum [(n_x)(x - \text{MEAN})^2] / \sum (n_x) = 43.4/112 = 0.39$$

Finally, by knowing that there is a relationship between the total number of crossings (CROSS) and the linear density (λ_l) of the form:

$$\lambda_l = \pi \text{CROSS}/(2 \text{TLA})$$

where TLA is the total length of the auxiliary transect lines (TLA = 5.6 km), we have

$$\lambda_l = \pi 104/(2 \times 5.6\text{km}) = 29.2 \, \text{km}^{-1} = 29.2 \, \text{km/km}^2$$

Table 3.3 Summary of the counts of crossings per segment.

Segment size (s)	MEAN of the number of crossings	VARIANCE of the number of crossings	VARIANCE/MEAN ratio (VMR)
100	1.86	0.59	0.32
50	0.93	0.39	0.42

These calculations show that the method of the auxiliary transect lines can be very useful to provide information for both linear density and distribution pattern, two very important characteristics of the distribution of linear features in the landscape.

3.6 A Study of Linear Features on the European Scale

A good example of a study that used these methods was that of van der Zanden and others[23] who used 250 m line transects in a sampling grid throughout Europe and recorded the number of crossings with linear features (Figures 3.23 and 3.24).

The distribution of all three types of linear features was found to be overdispersed (regularly distributed) as, for all cases, the variance was lower than the mean and the VARIANCE to MEAN ratio (VMR) was therefore always less than unity. For green lines VMR = 0.30/0.70 = 0.43, for ditches VMR = 0.26/0.57 = 0.46, and for grass margins VMR = 0.46/1.12 = 0.41. As the values for VMR were similar and lower than unity in all cases it was concluded that regularity was dominant in all linear features analyzed.

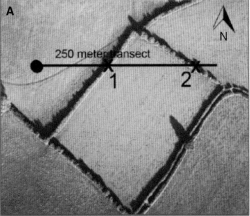

Figure 3.23 An overview of the grid used in transect-sampling linear features in Europe, with an example from southeastern Britain. *Source:* Van derZanden, E.H., Verburg, P.H., and Mucher, C.A. (2013) Modelling the spatial distribution of linear landscape elements in Europe. *Ecological Indicators*, 27, 125–136.

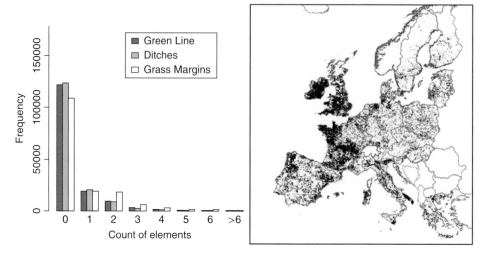

Figure 3.24 Results of the study showing the frequency of counts per transect for the three different types of linear features analyzed: green lines (hedge and tree lines), ditches, and grass margins (left). On the right, the spatial model of the results for hedge and tree lines showing higher densities in some regions where they are part of traditional agricultural landscapes in France, England, Wales, and Ireland or in northern Portugal and northwestern Spain. *Source:* Van derZanden, E.H., Verburg, P.H., and Mucher, C.A. (2013) Modelling the spatial distribution of linear landscape elements in Europe. *Ecological Indicators*, 27, 125–136.

3.7 The Topology of the Networks

After determining the two first characteristics of a linear network, the linear density and the spatial pattern, it is important to understand the topology of the network. Here we have to combine points and lines, as our definition of a network is a system of points (nodes) connected by lines (linkages). Many examples of networks can illustrate this definition. In transportation theory, nodes are generally viewed as cities connected by linkages (railway lines, roads), which are measured in distance or time units (Figure 3.25). Intersection nodes are simply the junction of intersecting corridors. From the organism perspective and nature conservation, cities are analogous to acceptable habitat and roads to ecological corridors.

The importance of the linkages between nodes can also be illustrated by the Florida Wildlife Corridor Initiative, aiming at establishing linkages (corridors) between nodes (areas of current and potential habitat for the Florida panther (*Puma concolor couguar*)) (Figure 3.26).

In spite of the many favorable results obtained, the value of corridors for conservation has been a subject of controversies in recent years due to their potential for the spread of invasive species, diseases, or disturbances.

Regardless of some controversies, the concept of ecological corridors has gained important recognition in various sectors, especially in land use planning (Figure 3.27)[24]. This is one of the basic principles of landscape ecology and has been used in landscape architecture and land-use planning.

We already know that the quantification of a network starts with the measurement of the density of nodes (number of nodes per unit area) and the linear density of the network (total length of lines per unit area) and the pattern of their distribution.

We should now consider another aspect of the network, its connectivity.

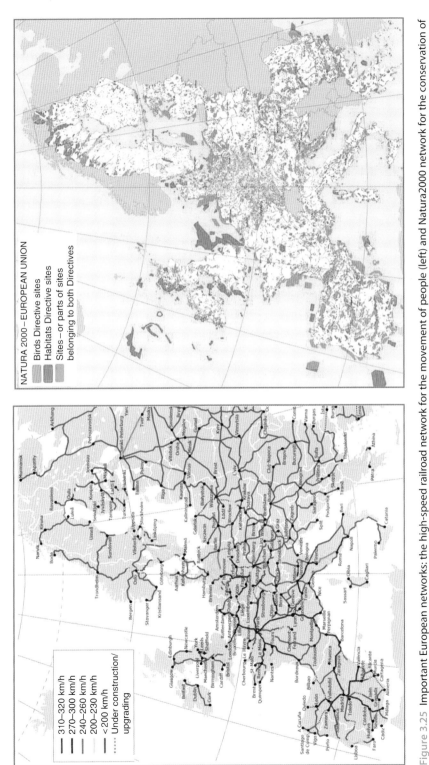

Figure 3.25 Important European networks: the high-speed railroad network for the movement of people (left) and Natura2000 network for the conservation of species and habitats (right). *Sources:* https://en.wikipedia.org/wiki/High-speed_rail_in_Europe (left), https://www.eea.europa.eu/data-and-maps/figures/natura-2000-birds-and-habitat-directives-8 (right).

Figure 3.26 The Florida panther (*Puma concolor couguar*) and the potential Florida panther corridor system connecting currently occupied habitat with large areas of potential habitat. *Sources:* SFWMD. GOV (above), the Florida Wildlife Corridor Initiative, http://floridawildlifecorridor.org/maps/ (below).

3.8 Network Connectivity

Connectivity is measured by the degree to which all nodes in a system are linked by corridors. The quantification of connectivity may use the index (ICON) that relates the observed number of links in the network (ONL) with the maximum number of links

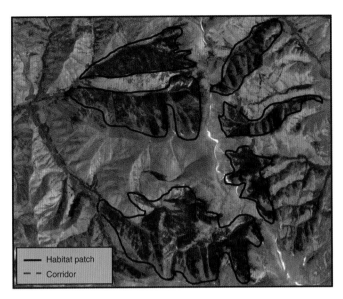

Figure 3.27 Landscape indicating forested habitat patches and connecting corridors in the Lost River Range of central Idaho. Lines show connectivity (linkages) between patches (nodes) of favorable habitat. *Source:* Google Images.

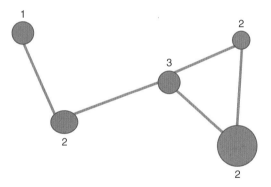

Figure 3.28 Network showing the degree of the different nodes.

(MNL) for a network with the same number of nodes (V) (Figure 3.28). The index[25] ICON is therefore calculated as

$$ICON = ONL/MNL = ONL/[3(V-2)]$$

The value $3(V-2)$ is the maximum number of links in a network assuming that all intersections between links are considered nodes. This index of connectivity ranges from 0 (isolated nodes) to 1 (all nodes connected to all other nodes). The number of linkages of a node (the degree of the node) is also a good indicator of the network pattern.

In this example we have five nodes ($V = 5$) connected by five links (ONL = 5). The connectivity index is thus

$$ICON = 5/[3(5-2)] = 5/9 = 0.56$$

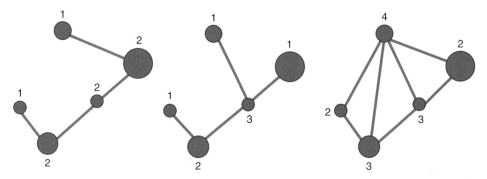

Figure 3.29 Linear, dendritic, and rectilinear networks with 5 nodes ($V = 5$). In this example, the maximum number of links is MNL = $3(V - 2) = 9$. In the first two networks the number of links is 4 (ONL = 4) and therefore we have a connectivity of ICON = $4/9 = 0.44$ whereas for the last network we have 7 links and a value of connectivity of ICON = $7/9 = 0.77$.

The network is only partially connected. The number of links at each node is in the image and indicates that more linkages could be possible.

Networks may approach three basic forms (topologies): linear (one railroad), dendritic (stream network), and rectilinear (hedgerow) (Figure 3.29).

Very simple linear networks generally have 2 linkages originating at each node and vertices have only 1 linkage. For dendritic networks, such as stream systems, nodes with a degree of 1 (vertices) and 3 are common. Rectilinear networks, common for road systems and hedgerows, typically have many nodes originating 3 and 4 linkages.

It is often of interest to maximize network connectivity, as in the case of evaluating alternative greenway networks and linkages[26,27]. However, the distance between nodes can also be taken into account. In this case the ratio between the number of linkages and their total length can provide a good way to evaluate alternative scenarios or to compare different landscapes. Also, it is possible in transportation theory to derive network topologies to solve basic problems. If we want all nodes to be minimally linked, the solution is a simple linear network, as shown in the example, where nodes are either extreme (degree one) or degree two. For a "traveling salesman" problem of a minimal loop network a solution is obtained for all nodes having a degree of two[28].

The use of representation of landscape features as nodes and links may be well demonstrated in the process of designing a network of reserves. In this case nodes represent important habitat patches and links may represent the shortest pathways among those patches. These problems can be solved in the framework of graph theory, as demonstrated in several studies since the pioneer work of Ricotta *et al.*[29]

These studies based on graph theory allow for many simulations of the effects of options on the reserve network. One of the common options is the establishment of the minimum spanning tree, which is the shortest path between all nodes that does not include cycles (paths of three or more nodes). These minimum spanning trees can then be assessed as landscape graphs for features as recruitment, dispersal flux, or traversability, and the sensitivity of node removal can be evaluated indicating their relative importance for the network, as shown in Figure 3.30.

Other studies follow related approaches[31] with the same objective of designing reserve networks based on species characteristics and connectivity.

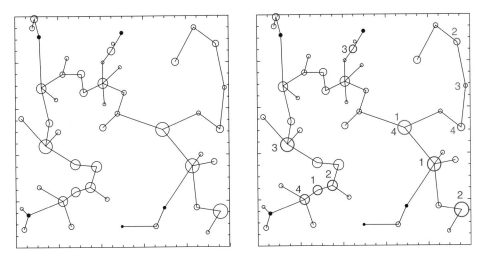

Figure 3.30 Left: hypothetical landscape with 50 circular habitat patches connected by a minimum spanning tree based on dispersal probabilities between patches. Right: results of a node removal exercise showing the most important patches (ranked 1–4) for recruitment (red), dispersal flux (blue), and traversability (green)[30].

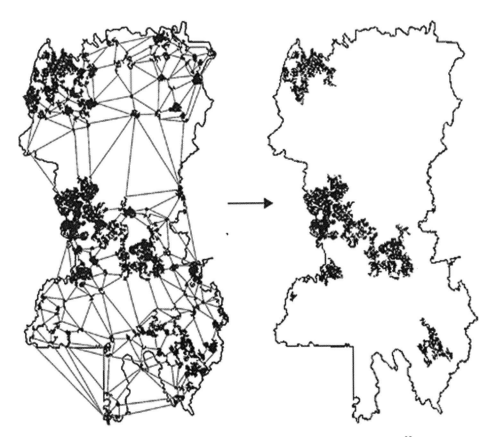

Figure 3.31 Design of a reserve network for the Vermillion area in Quebec, Canada[32]. The minimum planar graph (left) links all nodes (patches) with the maximum number of links and shortest length. The final reserve network proposed (right) aims at maintaining the maximum landscape connectivity for the species considered (brown creeper (*Certhia americana*), pileated woodpecker (*Dryocopus pileatus*), and American marten (*Martes americana*)).

This is the case of the design of a reserve network in the Vermillion area in the province of Quebec, Canada (Figure 3.31). In this study potential reserve patches were identified as nodes and a minimum planar graph was established with the maximum number of non-crossing links among nodes that are of shortest length. Then the importance of the patches was determined at multiple scales corresponding to the different dispersal abilities of the species to protect. The final reserve network was designed as the overlap of the patches identified as most important for landscape connectivity for each of the species.

These two studies illustrate different possibilities of using the approach of graph theory with nodes and links to design reserve networks in the landscape that take into account the ecological characteristics of the species to protect.

3.9   Connectivity Indices Based on Topological Distances Between Patches (Nodes)

Some relatively simple indices based on the topological distance between nodes are available and will be illustrated in this chapter. These indices are based on the work of Frank Harary on graph theory in the United States and have been developed and used in many different situations, such as in evaluating the connectivity in forested or grassland dominated landscapes, by many authors.

In Europe scientists such as Carlo Ricotta and Santiago Saura have been very influential in developing and promoting the use of indices based on graph theory and topological distances between nodes to evaluate connectivity in landscapes (Figure 3.32).

The topological distances between two nodes (d_{ij}) is simply the minimum number of links that connect them.

In Figure 3.33 we use the same example as in Figure 3.29 to illustrate the calculations of all the topological distances between all pairs of nodes and the calculation of one of the

Figure 3.32 Two influential scientists using graph theory in landscape analyses: Frank Harary (1921–2005), American mathematician and one of the "fathers" of graph theory (left), and Carlo Ricotta, Professor of Landscape Ecology at the University of Rome "La Sapienza" (right). *Sources:* New Mexico State University (http://newscenter.nmsu.edu/Articles/view/4538) (left), Carlo Riccota (right).

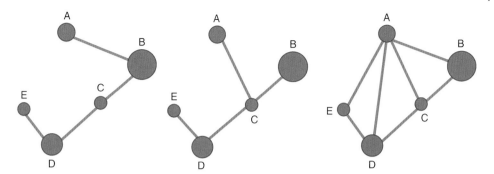

Figure 3.33 The three networks already presented in Figure 3.29 are used in the analysis of connectivity based on topological distances between nodes.

first indices of connectivity based on topological distances between nodes: the Harary index (HI) computed as the sum of the inverse values of the topological distance (minimum number of links) between every two nodes. If two nodes are not connected, their topological distance is infinity. The index HI can be simply computed by the equation:

$$HI = \sum\sum (1/d_{ij}) \quad \text{with } j > i$$

where d_{ij} is the topological distance (the minimum number of links) between different nodes i and j.

The calculations of the Harary index (HI) for the three networks represented in Figure 3.33 are now presented. The half-matrices with the topological distances between pairs of nodes are shown for the three networks, followed by the half-matrices with the inverse of the distances at the bottom, with the corresponding sum, the value of the Harary index (HI):

	A	B	C	D	E
A		1	2	3	4
B			1	2	3
C				1	2
D					1
E					

	A	B	C	D	E
A		2	1	2	3
B			1	2	3
C				1	2
D					1
E					

	A	B	C	D	E
A		1	1	1	1
B			1	2	2
C				1	2
D					1
E					

	A	B	C	D	E
A		1.00	0.50	0.33	0.25
B			1.00	0.50	0.33
C				1.00	0.50
D					1.00
E					

	A	B	C	D	E
A		0.50	1.00	0.50	0.33
B			1.00	0.50	0.33
C				1.00	0.50
D					1.00
E					

	A	B	C	D	E
A		1.00	1.00	1.00	1.00
B			1.00	0.50	0.50
C				1.00	0.50
D					1.00
E					

HI 6.42   **HI** 6.67   **HI** 8.50

The first example (left) shows a linear chain network where the nodes are increasingly distant from each other and HI = 6.42 (the minimum value for a linked network of 5 nodes); in the center the value of the connectivity index HI is slightly higher, HI = 6.67, and on the right a very connected network, HI = 8.50.

The Harary index has been used in many studies after it has been presented and described in pioneer works that quantify network connectivity of landscape mosaics based on graph theory[33]. However, after reviewing many of the existing connectivity indices, Saura and Pascual-Hortal[34,35] identified a number of shortcomings and proposed the integral index of connectivity (IIC). As with the Harary index, the IIC is also calculated from the topological distances between nodes (patches) but it includes attributes of the patches, typically their area, as it influences the estimated dispersal flux between different patches. In this approach the ecological system is modeled as habitat patches (nodes) and connections (links). Any loss or change of either nodes or their links results in a reduction of landscape connectivity for a particular species or process.

Two complementary metrics of landscape connectivity have been used in recent studies.

The first is the number of habitat components. A component is defined as a set of habitat patches with a connection between every two habitat patches within it. As the landscape becomes more connected, the number of components within the landscape decreases[36]. Also, a threshold distance must be specified for a link to be considered to exist. Obviously, as the threshold distance increases the number of components (non-connected groups of patches) decreases.

The second connectivity metric is the integral index of connectivity (IIC), based on graph theory, and is capable of combining many spatially explicit habitat variables and species-specific characteristics into a landscape connectivity value[37,38]. The equation for the IIC is

$$\text{IIC} = \left(1/\text{MAX}^2\right) \sum \sum \left[(a_i a_j)/(1 + d_{ij})\right] \text{ with } j > i$$

where a_i and a_j are the quantitative characteristic attributes of the node such as area (or habitat quality that may be relevant), d_{ij} is the topological distance or the number of links in the shortest path between patches i and j, and MAX is the maximum value of a given landscape attribute, for example the total area of the landscape or the total area of the patch type considered.

Using the networks shown in Figure 3.33 we can illustrate the calculation of IIC. The size of the patches (nodes) was established as

Patch	Size of patch (hectares)
A	2
B	5
C	1
D	3
E	1

To simplify calculations we consider MAX to be the total area of the patches considered, M = 12 hectares. We can now compute the values for each pair of nodes. For example, for the first network, for the pair of nodes A and B the sizes are 2 and 5 and the topological distance is 1. Therefore the corresponding value is $(2 \times 5)/(1 + 1) = 5.00$. We can now compute all the values for the half matrices ($j > i$), their sum, and the value of IIC (the sum of the elements of the half-matrix divided by MAX^2) for the three networks:

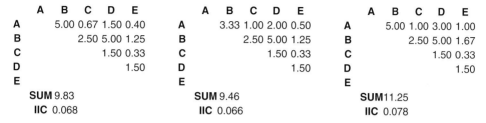

	A	B	C	D	E
A		5.00	0.67	1.50	0.40
B			2.50	5.00	1.25
C				1.50	0.33
D					1.50
E					

SUM 9.83
IIC 0.068

	A	B	C	D	E
A		3.33	1.00	2.00	0.50
B			2.50	5.00	1.25
C				1.50	0.33
D					1.50
E					

SUM 9.46
IIC 0.066

	A	B	C	D	E
A		5.00	1.00	3.00	1.00
B			2.50	5.00	1.67
C				1.50	0.33
D					1.50
E					

SUM 11.25
IIC 0.078

The results show again that the network at the right is more connected than those at the left and center.

The IIC has been successfully utilized in assessing landscape connectivity in many vegetation types from throughout the world[39]. Fourie and others used the IIC to assess connectivity in South African landscapes, considering only grassland patches or both grassland and abandoned agricultural lands in the analysis[40]. Due to the lack of information available on the dispersal distances for many species present, a range of dispersal distances were assessed including: 40, 100, 250, 500, and 1000 m. Obviously, when using larger threshold distances grassland patches were considered to be more connected (low number of components and higher values for the integral index of connectivity). When abandoned agricultural lands were included in the analysis the connectivity of the landscape increased (Figures 3.34 and 3.35).

The importance of each individual patch for the overall connectivity of the patch type can now be calculated by removing that patch and calculating the new IIC after removal of the patch (IIC_{after}). The percent reduction of the integral index of connectivity for the landscape is the connectivity value of the patch (dIIC):

$$dIIC(\%) = 100(IIC - IIC_{after})/IIC$$

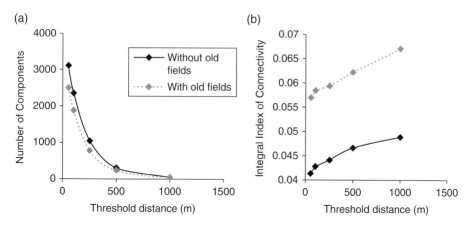

Figure 3.34 (a) Number of components and (b) integral index of connectivity of the grassland habitat patches of Mpumalanga, including and excluding abandoned croplands (old fields). *Source:* Fourie, L., Rouget, M., and Lötter, M. (2015) Landscape connectivity of the grassland biome in Mpumalanga, South Africa. *Australian Ecology*, 40, 67–76.

Figure 3.35 The Blyde River Canyon Nature Reserve, Panorama Route, Mpumalanga, South Africa. *Source:* https://commons.wikimedia.org/wiki/File:20131119_162559b.jpg.

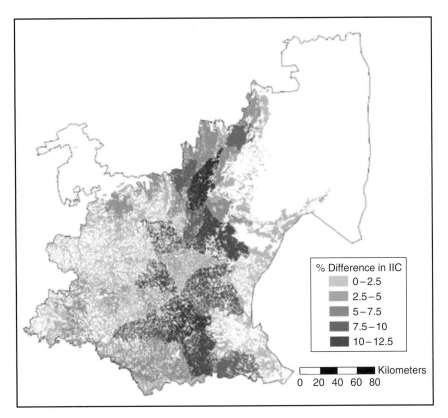

Figure 3.36 The percent differences in the value of the integral index of connectivity (IIC) expressed as percentages dIIC(%) for the grassland patches of Mpumalanga showing that the importance of the individual patches for the connectivity of that habitat in the landscape is geographically highly variable. *Source:* Fourie, L., Rouget, M., and Lötter, M. (2015) Landscape connectivity of the grassland biome in Mpumalanga, South Africa. *Australian Ecology*, 40, 67–76.

where IIC is the overall index value when all nodes are present in the landscape and IIC$_{after}$ is the overall index value after the removal of the specific habitat patch[41,42].

The results for the grasslands patches of Mpumalanga[43] can illustrate the situation (Figure 3.36).

The calculation of the values of the integral index of connectivity (IIC) for a certain patch type in the landscape for individual patches of that type (dIIC) can be performed using computer programs that have been incorporated in the system Conefor, making it readily usable by landscape ecologists and planners.[44]

Although the focus of this chapter is on linear elements and networks more developments on using information on patch attributes as patch size and shape and interactions between patches are presented in the next chapter.

Key Points

- Linear features, which can be line corridors or strip corridors (depending on scale and the objective of the analysis), influence many ecological characteristics and processes in the landscape. Their function can be of conduit, barrier and/or filter, habitat, source and/or sink.
- Linear elements can be represented by single lines, with no width, and are always characterized by length and curvilinearity. The quantification of length depends on the unit of measurement and therefore on the scale of the analysis. The concept of the fractal dimension D is used to measure the curvilinearity of a line and helps to explain how length changes with scale.
- A perfectly straight line has a value D of 1 ($D = 1$) and the measured length is always independent of the size of the ruler or the scale of the analysis. As the curvilinearity of a line increases, the value D will range between 1 and 2 ($1 \leq D < 2$) and the length of the line depends upon the ruler used to measure it and on the scale of the analysis.
- The box-counting method uses quadrats of different sides to measure the curvilinearity of lines. The length of the line will be the number of quadrats filling the line. As in fractal dimensions, the size of a straight line is independent of the quadrat side and as curvilinearity increases the length of the line depends upon the quadrat side and on the scale of the analysis.
- Linear networks can be described by their density and spatial distribution. They affect ecological processes depending on the type of network and the process considered.
- Density, the most important characteristic of a network of lines, measures the abundance of lines and/or corridors in an area, and is defined as the total length of linear elements per unit area.
- Several methods can be used to determine the spatial pattern or distribution of a network of lines. Methods are based on quadrats, to estimate the fractal dimension of the network of lines, or on distances, to calculate the nearest-neighbor distance between lines.
- The topology of the networks combines points (nodes) and lines (linkages), and is described by the density of nodes (number of nodes per unit area), by the linear density of the network (total length of lines per unit area), and by the pattern of their distribution.

- Connectivity, another characteristic of networks, is measured by the degree to which all nodes (points) in a system are linked by corridors (lines). Depending on their connectivity, networks may approach three basic forms: linear, dendritic, and/or rectilinear.
- Quantification of connectivity may use simple indices relating the number of nodes and links or indices based on graph theory that use the concept of topological distances between nodes and may include attributes of nodes (as patch sizes).

Endnotes

1 Forman, R.T.T. (1995) *Land Mosaics: The Ecology of Landscapes and Regions*, Cambridge University Press, Cambridge, MA.
2 Borda-de-Água, L., Grilo, C., and Pereira, H.M. (2014) Modeling the impact of road mortality on barn owl (*Tyto alba*) populations using age-structured models. *Ecological Modelling*, 276, 29–37.
3 Forman, R.T.T, Sperling, D., Bissonette, J.A., *et al.* (2003) *Road Ecology: Science and Solutions*, Island Press, Washington, DC.
4 Forman, R.T.T. (1995) *Land Mosaics: The Ecology of Landscapes and Regions*, Cambridge University Press, Cambridge, MA.
5 Little, C.E. (1990) *Greenways for America*. Johns Hopkins University Press, Baltimore, MD.
6 Smith, D.S. and Hellmund, P.C. (eds) (1993) *Ecology of Greenways: Design and Function of Linear Conservation Areas*, University of Minnesota Press, Minneapolis, MN.
7 Fabos, J.G. and Ahern, J. (eds) (1996) *Greenways: The Beginning of an International Movement*, Elsevier, Amsterdam.
8 Ahern, J.F. (2002) *Greenways as Strategic Landscape Planning: Theory and Application*, Bepress, The Netherlands, http://works.bepress.com/ahern_jack/7.
9 Pennycuick, C.J. and Kline, N.C. (1986) Units of measurement for fractal extent, applied to the coastal distribution of bald eagle nests in the Aleutian islands, Alaska. *Oecologia*, 68, 254–258.
10 Mandelbrot, B.B. (1967) How long is the coast of Britain? Statistical self-similarity and fractional dimension. *Science*, 156, 636–638.
11 Fractal Foundation. Inspiring Interest in science, Math & Art, Online Course on Fractal Dimension of Coastlines. http://fractalfoundation.org/OFC/OFC-10-4.html
12 Dingman, S.L. (1994) *Physical Hydrology*, Prentice Hall, Upper Saddle River, NJ.
13 Sappington, J.M., Longshore, K.M., and Thompson, D.B. (2007) Quantifying landscape ruggedness for animal habitat analysis: A case study ssing bighorn sheep in the Mojave Desert. *Journal of Wildlife Management*, 71, 1419–1426.
14 McKinney, T., Boe, S.R., and deVos, J.C. (2003) GIS-based evaluation of escape terrain and desert bighorn sheep populations in Arizona. *Wildlife Society Bulletin*, 31, 1229–1236.
15 Brocke, R.H., O'Pezio, J.P., and Gustafson, K.A. (1990) A forest management scheme mitigating impacts of road networks on sensitive wildlife species, in *Is Forest Fragmentation a Management Issue in the Northeast?* (eds R.M. DeGraaf and W.M. Healey), USDA Forest Service General Technical Report NE-140, Northeastern Forest Experiment Station, Radnor, PA.

16 Wakkinen, W.L. and Kasworm, W.F. (1997) *Grizzly Bear and Road Density Relationships in the Selkirk and Cabinet-Yaak Recovery Zones,* USDI Fish and Wildlife Service Report, Missoula, MT.

17 Wakkinen, W.L. and Kasworm, W.F. (1997) *Grizzly Bear and Road Density Relationships in the Selkirk and Cabinet-Yaak Recovery Zones,* USDI Fish and Wildlife Service Report, Missoula, MT.

18 Boulanger J. and Stenhouse, G.B. (2014) The impact of roads on the demography of grizzly bears in Alberta. Open-access article. *PLoS ONE,* 9 (12), e115535.

19 Davis, J.C. (1973) *Statistics and Data Analysis in Geology,* John Wiley & Sons, Inc., New York, NY.

20 Davis, J.C. (1973) *Statistics and Data Analysis in Geology,* John Wiley & Sons, Inc., New York, NY.

21 Sellman, P.V. and Dingman, S.L. (1970) Prediction of stream frequency from maps. *Journal of Terramechanics,* 7, 101–115.

22 Davis, J.C. (1973) *Statistics and Data Analysis in Geology,* John Wiley & Sons, Inc., New York, NY.

23 Van der Zanden, E.H., Verburg, P.H., and Mucher, C.A. (2013) Modelling the spatial distribution of linear landscape elements in Europe. *Ecological Indicators,* 27, 125–136.

24 Dramstad, W. Olson, J., and Forman, R. (1996) *Landscape Ecology Principles in Landscape Architecture and Land Use Planning,* Island Press, Washington, DC.

25 Forman, R.T.T. and Godron, M. (1986) *Landscape Ecology,* John Wiley & Sons, Inc., New York, NY.

26 Bueno, J.A., Tsihrintzis, V.A., and Alvarez, L. (1995) South Florida greenways: a conceptual framework for the ecological connectivity of the region. *Landscape and Urban Planning,* 33, 247–266.

27 Linehan, J., Gross, M., and Finn, J. (1995) Greenway planning: developing a landscape ecological network approach. *Landscape and Urban Planning,* 33, 179–193.

28 Haggett, P. and Chorley, R.J. (1972) *Network Analysis in Geography,* Edward Arnold, London.

29 Ricotta, C., Stanisci, A., Avena, G.C., and Blasi, C. (2000) Quantifying the network connectivity of landscape mosaics: a graph-theoretic approach. *Community Ecology,* 1, 89–94.

30 Urban, D. and Keitt, T. (2001) Landscape connectivity: a graph theoretic perspective. *Ecology,* 82, 1205–1218.

31 Orestes, J., Gaston, K., and Pinto, L.S. (2005) Connectivity in priority area selection for conservation. *Environmental Modeling and Assessment,* 10, 183–192.

32 James, P., Rayfield, B., Fortin, M.J., *et al.* (2005) Reserve network design combining spatial graph theory and species´ spatial requirements. *Geomatica,* 59 (3), 323–333.

33 Ricotta, C., Stanisci, A., Avena, G.C., and Blasi, C. (2000) Quantifying the network connectivity of landscape mosaics: a graph-theoretic approach. *Community Ecology,* 1, 89–94.

34 Pascual-Hortal, L. and Saura, S. (2006) Comparison and development of new graph-based landscape connectivity indices: towards the priorization of habitat patches and corridors for conservation. *Landscape Ecology,* 21, 959–967.

35 Saura, S. and Pascual-Hortal, L. (2007) A new habitat availability index to integrate connectivity in landscape conservation planning: comparison with existing indices and application to a case study. *Landscape and Urban Planning,* 83, 91–103.

36 Saura, S. and Pascual-Hortal, L. (2007) A new habitat availability index to integrate connectivity in landscape conservation planning: comparison with existing indices and application to a case study. *Landscape and Urban Planning*, 83, 91–103.

37 Calabrese J.M. and Fagan W.F. (2004) A comparison-shopper's guide to connectivity metrics. *Frontiers in Ecology and the Environment*, 2, 529–536.

38 Fourie, L., Rouget, M., and Lötter, M. (2015) Landscape connectivity of the grassland biome in Mpumalanga, South Africa. *Australian Ecology*, 40, 67–76.

39 Conefor - Quantifying the importance of habitat patches and links for landscape connectivity, http://www.conefor.org/index.html (Accessed June 1 2017).

40 Fourie, L., Rouget, M., and Lötter, M. (2015) Landscape connectivity of the grassland biome in Mpumalanga, South Africa. *Australian Ecology*, 40, 67–76.

41 Pascual-Hortal, L. and Saura, S. (2006) Comparison and development of new graph-based landscape connectivity indices: towards the priorization of habitat patches and corridors for conservation. *Landscape Ecology*, 21, 959–967.

42 Saura, S. and Pascual-Hortal, L. (2007) A new habitat availability index to integrate connectivity in landscape conservation planning: comparison with existing indices and application to a case study. *Landscape and Urban Planning*, 83, 91–103.

43 Fourie, L., Rouget, M., and Lötter, M. (2015) Landscape connectivity of the grassland biome in Mpumalanga, South Africa. *Australian Ecology*, 40, 67–76.

44 Saura, S. and Torné, J. (2009) Conefor Sensinode 2.2: a software package for quantifying the importance of habitat patches for landscape connectivity. *Environmental Modelling and Software*, 24, 135–139.

4

Patches and Their Interactions

4.1   The Importance of Patch Size for Species Diversity

Patches are defined as relatively homogeneous nonlinear areas that differ from their surroundings[1]. Nevertheless, the description and classification of patches should focus on characteristics that have relevant ecological implications for the organism, or the process, of interest.

We will start by describing the characteristics of an individual patch and their influence on processes before discussing the interactions between patches of the same type in the landscape. Three basic characteristics need to be used in the description of a single patch: size (area), edge (perimeter), and shape (typically relating perimeter and area).

Patch size is generally the first characteristic of a patch to be measured and considered. In fact, patch size is well known to be of great importance for organisms but some authors[2] also indicate that the effect of patch size on ecosystem processes (hydrological regimes, mineral nutrient cycles, radiation balance, wind patterns, and soil movement) is just as great. Furthermore, these processes also influence species patterns[1].

Species diversity is dependent on patch size and the concept of the species–area relationship can be used to represent it graphically. The rule that the number of species found increases with the increase of the area sampled has more evidence to support it than any other rule about species diversity[3].

Species–area relationships were probably first noticed, before any other diversity pattern, by H.C. Watson, in his works between 1847 and 1859, the distribution of plant species in Great Britain[4] (Table 4.1), as indicated by many other authors such as Williams, in 1964, in his influential book on *Patterns in the Balance of Nature*[5]. We will therefore use the example from Watson to illustrate the relationship between number of plant species and area (Figure 4.1).

Many other plant ecologists have been using species–area curves for many purposes. For example, species–area curves developed while sampling within a plant community were also used for a long time to determine the minimum area for an adequate sampling protocol[6]. This minimum area is the point at which an abrupt change in slope takes place and the curve tends to level out and has been considered an indication of how large a patch must be to express its composition.

Examples from these relationships between number of species and area have also been applied to island species diversity. In their pioneer work on island biogeography,

Applied Landscape Ecology, First Edition. Francisco Castro Rego, Stephen C. Bunting, Eva Kristina Strand and Paulo Godinho-Ferreira.
© 2019 John Wiley & Sons Ltd. Published 2019 by John Wiley & Sons Ltd.

Table 4.1 The original table for the relationship area/number of plant species provided by Watson for Britain.

Area	Square miles	Number of plant species
All Britain	87 412	1425
England	57 812	1350
South Britain	38 474	1280
Province of Thames	7007	1051
South Thames	2316	972
County of Surrey	760	840
Part of N. Surrey	60	660
Ten square miles	10	600
One square mile	1	400

Adapted from Watson H.C. (1847–1859) *Cybele Britannica: Or British Plants and Their Geographical Relations*, Longman, London.

Figure 4.1 The English botanist and plant geographer Hewett C. Watson (1804–1881), who studied and published (1847–1859) the flora of the counties and provinces of Britain. *Source:* https://upload.wikimedia.org/wikipedia/commons/d/d4/Hewett_Cottrell_Watson.jpg (Public domain), Watson, H.C. (1847–1859) *Cybele Britannica: Or British Plants and Their Geographical Relations*, Longman, London.

MacArthur and Wilson[7] observed that the number of amphibians and reptiles of the West Indian islands could be plotted against the area of the island along a straight line in a log–log plot. Animal ecologists have settled in that standard way to plot species–area curves for analysis. The equation that relates the number of species on an island or patch (S) and its area (A) is considered to have generally the form:

$$S = cA^z$$

or the equivalent linear relationship:

$$\log(S) = \log(c) + z \log(A)$$

where c and z are parameters that can be fitted to the various situations.

Using the above example for the number of plant species for various areas inside Great Britain we can use a log–log plot to represent the relationship (Figure 4.2). In this case we use, for an easier understanding, the logarithms with base 10. On the horizontal axis of such a log–log graph, an area of 1 square mile will have a value of 0 and an area of 10^6 square miles will have a value of 6. Similarly, for the species axis, a number of 10^3 species, close to what was observed in the province of Thames, would yield a value of 3 in the vertical axis.

A linear equation is shown to have a very good fit ($R^2 = 0.989$) and is expressed by

$$\log_{10}(\text{number of plant species}) = 2.632 + 0.105 \log_{10}(\text{area in square miles})$$

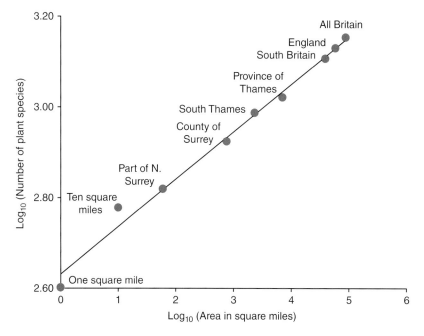

Figure 4.2 The log–log plot (base 10) applied on the plant species/area data of Watson for Great Britain. *Source:* Watson, H.C. (1847–1859) *Cybele Britannica: Or British Plants and Their Geographical Relations,* Longman, London.

It should be noted here that the units of area are not important for the determination of z. If the area was expressed is km^2 (A') instead of square miles (A) we would have

$$A' = bA \quad \text{or} \quad A = A'/b$$

where b is a constant (conversion factor $2.59\ km^2$/square mile).
In this case we would have

$$\log(S) = \log(c) + z \log(A'/b) = \log(c) - z \log(b) + z \log(A')$$

The intercept would be different but the slope of the line (z) would be the same, showing that the value of z is independent of the units of area. We will therefore keep the graphs in the area units of the original papers (square miles).

Since the pioneer work of Watson many other empirical examples of log–log plots of species–area relationships for plants and animals have been established and the integration of spatial scales analyzed by some authors (Figure 4.3). A very good summary of those studies and the values for c and z reported in the literature for various plant and animal groups at different spatial scales was provided by Rosenzweig[8].

At the landscape level the observed pattern is also the increase in the number of species with the increase of patch size. This pattern is considered to be primarily due to the increase in the number of species that require larger interior habitats, that is species with larger home ranges. These "interior species" that have higher area requirements and require large habitat patches are generally those with larger body sizes. There have been

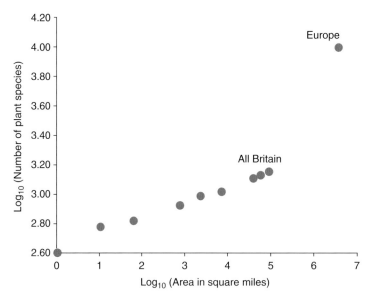

Figure 4.3 Graph showing the integration of the species–area curve developed for Britain in the global curve of plants of Europe using data from Watson[9], who indicated at that time around 10 thousand different plant species in the 3.65 million square miles of the area of Europe. Notice the different slopes in the two different portions of the plot (two different spatial scales), indicating that the slope of the species–area curve amongst separate biotas is much steeper than within biotas. *Source:* Williams, C.B. (1964) *Patterns in the Balance of Nature*, Academic Press, London.

various attempts to establish empirical relationships between home range size and body size for several animal groups.

A good illustration of these relationships is the simple equation, almost linear, proposed by Harestad and Bunnell[10] to relate home range size and body size for large mammalian herbivores:

$$\text{Home range (ha)} = 3.2 \, \text{body mass (kg)}^{0.998}$$

It is easy to understand that home ranges are related to body size and that home ranges also differ from birds to terrestrial animals and, among these, between herbivores and carnivores (Table 4.2). It is clear that a carnivore will always have a larger home range than a herbivore of the same size and that fast-moving animals, such as birds, have larger home ranges than equivalent, terrestrial animals. Other factors, such as the habitat's productivity and population density, can also affect home range size. This table with indications of home range estimates for various animal species can illustrate these relationships.

Table 4.2 Examples of home range estimates related to body mass for a number of vertebrate species.

Type/common name	Scientific name	Body mass (kg)	Home range (ha)
Terrestrial herbivores			
Meadow vole	*Microtus pennsylvanicus*	0.04	0.01
Snowshoe hare	*Lepus americanus*	2	6
White-tailed deer	*Odocoileus virginianus*	90	200
Red deer	*Cervus elaphus*	200	500
Elk	*Cervus canadiensis*	300	1300
Moose	*Alces alces*	400	1600
Terrestrial carnivores			
Stoat	*Mustela erminea*	0.09	20
American marten	*Martes americana*	1	200
Red fox	*Vulpes vulpes*	6	400
Grey wolf	*Canis lupus*	40	6000
Brown bear	*Ursus arctos*	200	9000
Birds			
Marsh tit	*Parus palustris*	0.01	2
Common blackbird	*Turdus merula*	0.1	40
Common buzzard	*Buteo buteo*	1.0	200
Common raven	*Corvus corax*	1.5	900

Data from various sources: (1) Harestad, A.S. and Bunnell, F.L. (1979) Home range and body weight – a reevaluation. *Ecology*, 60, 389–402; (2) Jorge, J.L.T. (1986) *Manual para el Censo de los Vertebrados Terrestres*, Editorial Raices, Madrid; and (3) Lindstedt, S.L., Miller, B.J., and Buskirk, S.W. (1986) Home range, time and body size in mammals. *Ecology*, 67, 413–418.

The combination of the home range of the species with the size of the patch and its habitat quality determines the carrying capacity (K) or the maximum local size of the population of that species within that patch. This could then be related to the probability of local extinction of that population using existing models such as the one proposed by MacArthur[11] describing demographic stochasticity.

The analysis of the probabilities of local extinction related to population size led to the use of the concept of a minimum viable population (MVP), often defined for any given species in any given habitat as the smallest isolated population having a very good chance (99%) of remaining extant for a long period (1000 years), despite the foreseeable effects of demographic, environmental and genetic stochasticity, and natural catastrophes[12].

Difficulties in the estimate of a universal number for the minimum viable population (MVP) arise from the various approaches applied and therefore different values have been proposed. Barbault and Sastrapradja[13], for example, suggest that demographic stochasticity poses a real threat to survival only when population size is very small, say fewer than 50 individuals, while Seal[14], based on various studies with vertebrate species in captivity, proposes that 250–500 individuals would be a reasonable estimate of MVP and Forman[15] considers a "few hundred to a few thousand" as the current overall hypothesis of MVP for natural populations. Despite these differences in estimates of MVP, all authors agree that the probability of extinction increases with decreasing population size and that it is controlled by the home range of the species and the size and quality of the patch of habitat.

4.2 The Importance of Patch Edge and Shape

Besides patch size, patch perimeter or edge, the boundary between adjacent patch types of different class has many important consequences for organisms. The emphasis on the importance of patch perimeters for some organisms (edge species) led to the oversimplified conclusion that land management sought to maximize edge in order to create the greatest amount of desirable habitat as possible for wildlife species. However, it was soon recognized that this approach favored creation of habitat for edge species but disfavored habitat for interior species. The need to recognize the balance between edge and interior habitat has been recently emphasized.

The simultaneous consideration of patch area and perimeter and the balance between interior and edge species is generally done by the analysis of patch shape. In fact, the shape of an individual patch has many consequences related to its ability to provide habitat for organisms and influences many landscape processes. Patch shape has been shown to influence processes as the microclimate of the patch, the movement of water, energy and materials, the spread of disturbances, the recruitment of plants[16], or the foraging strategies[17] and migration[18] of animals.

From an organism perspective the shape of a patch is important when determining how much of a patch is available as habitat for a given species. For example, some bird species are adversely affected by predation, competition, brood parasitism, and perhaps other factors along forest edges. For such interior species, requiring a certain distance from the patch boundary, the area beyond that distance from the edge (core area) better characterizes a patch than the total patch area[19]. The same concept of core area has also been applied in the context of home range[20], to denote central areas of consistent or intense use.

One of the first examples of the effect of patch size and shape on species diversity is the study of avian distribution patterns in oak woodlots of different sizes in central New Jersey by Forman and others[21] (Figure 4.4).

In evaluating how the size of the forest patch might affect the number of bird species, the authors found that herbivores and omnivores did not show a very sharp increase in oak patches above 2 to 3 hectares in size whereas insect feeders increased much more with patch size and vertebrate predators occurred only in the largest forest patches[22] (Figure 4.5). As the feeding territory size requirements of birds increase along a diet gradient from herbivores to insect feeders to vertebrate predators[23], herbivores and omnivores are typically edge species or species that occur indifferently in the edge and interior of the forest patch, whereas insect feeders and vertebrate predators are generally interior species (those that require large forest patches and are primarily away from the periphery).

The edge to interior ratio decreases with increased patch size. Therefore the effect of the increase in patch size on species diversity is often confounded with that of the edge/interior ratio. Nevertheless, it is important to notice that it is the influence of both patch size and shape with the home range of the species that is the determinant of species diversity. Because of their implications for organisms and processes, patch size, edge, and shape are all important characteristics to measure as they affect ecological processes and species diversity.

We saw before, in the chapter on linear elements, that the measurement of linear features depends on the measurement unit or the scale of the analysis. The same applies directly for the measurement of the perimeter of a patch where a fractal dimension can also be determined, ranging from values closer to unity for straight perimeters to

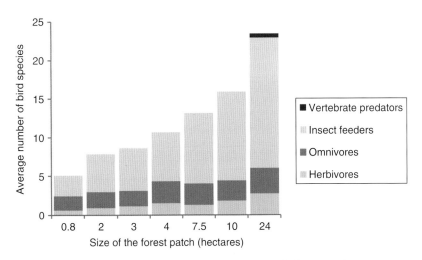

Figure 4.4 The number of bird species in relation to size of a forest patch in New Jersey, showing that the number of bird species increases especially with insect feeders that are primarily interior species. Adapted from Forman, R.T.T., Galli, A.E., and Leck, C.F. (1976) Forest size and avian diversity in New Jersey woodlots with some land use implications. *Oecologia*, 26, 1–8.

Figure 4.5 The Norvin Green State Forest in New Jersey where oaks dominate. The red shouldered hawk (*Buteo lineatus*) is a carnivore (vertebrate predator) and an interior bird species that is present only in the largest forest patches in order to obtain sufficient food (mice and snakes). *Source*: http://www. wikiwand.com/en/Northeastern_coastal_forests (left), https://upload.wikimedia.org/wikipedia/commons/d/d3/Red-shouldered_hawk_(Buteo_lineatus).jpg (right).

closer to 2 for very convoluted perimeters. The determination of the size of a patch (or a home range), which seems to be a straightforward procedure, is also dependent on the precision of the measurement unit.

4.3   The Measurement of Patch Size and Perimeter

The measurement of patch size is relatively simple as patches are defined as groups of raster cells (counts of cells multiplied by the area of a unit cell gives the area of the patch) or are given by the coordinates of the corners of the polygon (Figure 4.6). However, in some cases, these polygons have to be constructed. This is the case, for instance, when estimating the home range of an animal (polygon) based on locations of individual sightings (points). In this case one of the simplest and very intuitive approaches is the minimum convex polygon where the peripheral locations of an animal are connected in such a way that the internal angles of the polygon generated do not exceed 180°[24].

It is convenient, at this point, to exemplify how area and perimeter are determined when patches are represented as polygons, as in a vector-type representation in geographic information systems (GISs). In any case, both for patches and home ranges, the periphery of the polygon is represented by a series of n Cartesian coordinates X_i, Y_i), such as those generated by a digitizer.

Using pairs of these points we can construct a series of trapezoids that, starting from an arbitrary point and moving clockwise around the polygon, allow determination of the area (A) by the use of the expression

$$A = \Sigma[(X_{i+1} - X_i)(Y_{i+1} + Y_i)/2]$$

where X_i and Y_i are the Cartesian coordinates for point i.

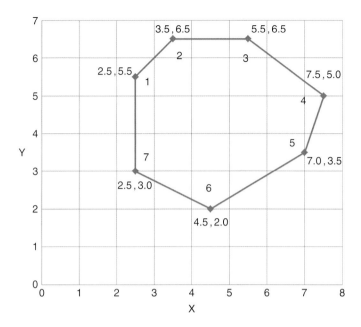

Figure 4.6 Artificial polygon used to illustrate the calculations of area and perimeter based on the X and Y coordinates of the 7 points of the periphery.

Using the same points and the same sequence we can also calculate the Euclidean distance between those points. The sum of these distances is the perimeter (P) of the patch:

$$P = \Sigma \left[(Y_{i+1} - Y_i)^2 + (X_{i+1} - X_i)^2 \right]^{1/2}$$

The calculations to compute the area (A in area units) and the perimeter (P in length units) of the example patch shown in Figure 4.6 are shown in Table 4.3.

This is a simple way to illustrate the calculations done with GIS systems for area and perimeter. It is important to remember again that the precision of the vector-type

Table 4.3 Calculations of the area and perimeter of a polygon based on the X and Y coordinates of the vertices (i) shown in Figure 4.6. The distance between point i and point i+1 equals d_{i+1}.

i	X_i	Y_i	$X_{i+1} - X_i$ (a)	$(Y_{i+1} + Y_i)/2$ (b)	(a)(b)	$Y_{i+1} - Y_i$ (c)	$(a^2 + c^2)$	d_{i+1} $(a^2 + c^2)^{1/2}$
1	2.5	5.5	1.0	6.00	6.00	1.0	2.0	1.41
2	3.5	6.5	2.0	6.50	13.00	0.0	4.0	2.00
3	5.5	6.5	2.0	5.75	11.50	-1.5	6.3	2.50
4	7.5	5.0	-0.5	4.25	-2.12	-1.5	2.5	1.58
5	7.0	3.5	-2.5	2.75	-6.88	-1.5	8.5	2.92
6	4.5	2.0	-2.0	2.50	-5.00	1.0	5.0	2.24
7	2.5	3.0	0.0	4.25	0.00	2.5	6.3	2.50
1	2.5	5.5						
					Area = 16.50			Perimeter = 15.15

representation of a patch is dependent upon the number and the distance between the points used to define the periphery, and that these have an influence on the final results.

In the previous analysis we studied methods to measure patch area and perimeter, and found that the results of measurements on complex shapes are generally scale-dependent. This may lead us to the conclusion that there is no single scale that best describes the shape of a patch. This is not necessarily correct from an organism-centered perspective. In fact, "the animal is the caliper" and the extent of its home range can be used as its caliper. For instance, in the example shown in the previous chapter the effective length of coastline available to bald eagles could be estimated using the average internest distance as the length unit. Similarly, the estimates of home range can provide a useful guide on the best area units to use in the analysis of landscapes for any given species. Consequently, there may be a "best" scale from the organism-centered perspective but the scale will change from one organism to another.

4.4 Quantifying Patch Shape

With the measurements of area (A) and perimeter (P) we can now start quantifying the shape of an individual patch. This shape may vary from very simple forms, such as the circle for vector representations and as the square for raster representations, to very complex and sinuous forms. For vector representations the overall shape of a patch is often evaluated in comparison with a circular standard since the circle is the shape that shows the lowest possible perimeter (P) for a given patch area (A).

One of the most commonly used measures is the shape index (SHAPE) that compares the actual perimeter (P) with the perimeter of a circle (P_c) with the area (A) of the patch. The perimeter of a circle (P_c) with area A is computed by using the equations for the perimeter and area of a circle:

$$P_c = 2\pi r \text{ and } A = \pi r^2$$

As in a circle $r = (A/\pi)^{1/2}$ we can compute P_c as

$$P_c = 2\pi (A/\pi)^{1/2} = 2(\pi A)^{1/2}$$

and the SHAPE index is computed from the equation

$$\text{SHAPE} = P/P_c = P/\left[2(\pi A)^{1/2}\right]$$

SHAPE is equal to 1 for a circle and increases with increasing complexity of the patch, that is, more perimeter length for the same area.

Another similar index (SHAPE2) can be computed as

$$\text{SHAPE2} = 1 + \ln(P/\pi) - (1/2)\ln(4A/\pi)$$

As with SHAPE, SHAPE2 also yields the value 1 for circular shapes and increases as shapes become more complex and convoluted. As an example of the polygon above where $A = 16.50$ area units and $P = 15.15$ length units we have

$$\text{SHAPE} = 1.05 \text{ and } \text{SHAPE2} = 1.05$$

These two indices, close to unity, indicate a very simple shape. However, they may provide more information if they are considered in relation to other values obtained with the same methods in different places or times.

The calculations presented up to this point apply for vector representations of patches. However, the same indices may be applied with minor modifications to raster representations of patches.

Using the example of the previous polygon we transform the vector representation to a raster representation, defining patch cells as those where at least half of the cell is occupied by the patch. In this case we can distinguish between edge cells, those adjacent to the perimeter line, and interior cells, those that do not have sides as a perimeter (Figure 4.7).

The total area of a raster representation of a patch can be determined by counting the cells occupied by the patch and multiplying it by the area of a single cell. With this type of representation the precision of the area estimate is dependent upon the spatial resolution (grain or cell size) used to depict the patch.

From the raster representation above we can now quantify the various characteristics of the patch. Now area (A) is measured as having 16 cells and the perimeter (P) as 20 units (cell sides). It is obvious that the measurements do not coincide with those made with the vector representation, and that is particularly significant in the measurement of the perimeter.

For raster representations the two indices already calculated can also be used with some modifications, as the reference unit, the shape that has a minimum perimeter for any given area, is not a circle but a square. In this case the minimum perimeter of a patch with area A would be obtained by a square of side $L = A^{1/2}$ and perimeter, $P = 4L = 4A^{1/2}$.

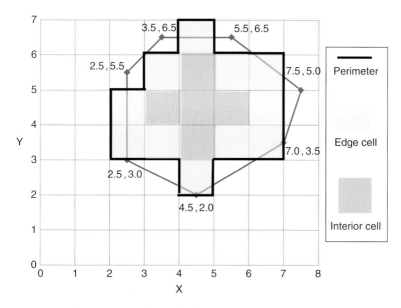

Figure 4.7 The same example patch of Figure 4.6 represented as a vector (polygon with perimeter in blue) and as a raster (cells with perimeters in black). Edge and interior (core) cells are in yellow and green, respectively.

The equations are

$$\text{SHAPE} = P \Big/ \left[4(A)^{1/2}\right]$$

$$\text{SHAPE2} = 1 + \ln (P/4) - (1/2) \ln (A)$$

For our example with $A = 16$ and $P = 20$, the computed indices would be

$$\text{SHAPE} = 1.25$$

and

$$\text{SHAPE2} = 1.22$$

These values are higher than those obtained with the vector representation of the patch, showing how methods can influence the results and highlighting the fact that the main importance of these values is in their comparison with standards or in comparison with other patches in time or space as measured by the same method.

In all cases, the indices approach 1 for simple shapes as that of the standard shape (circle or square), which has the minimum perimeter for a given patch area. The SHAPE indices increase as the shape becomes more complex and have no theoretical upper limit, as the maximum values are sensitive to the vector length used to display the patch.

4.5   An Example for the Use of Perimeter–Area Relationships

An example of the use of these relationships between perimeter and area can be provided by the shapes of the areas burned by wildfires (Figure 4.8). In this case the reference is,

Figure 4.8 Two alternative hypotheses for the shapes of fires: the shape of the fire remains simple at all sizes as the fire spreads in all directions with similar speed (left) or wildfire spread faces increasing obstacles that are different in the different directions, creating shapes that are increasingly complex with increasing size of the area burned (right). *Sources*: J.M.C. Pereira and the Portuguese Forest Services.

again, the circle and the hypothesis would be that, if there were no obstacles to their progression, fires would maintain simple shapes, from circular to elliptical, as expected in many theoretical models. In the alternative hypothesis, and due to the variations in the landscape factors, fuel, topography, wind, and the firefighting activities, the final shapes of the fires would get increasingly complex with increasing area.

These hypotheses were tested in northeastern Portugal, in Trás-os-Montes, using data from areas burned from 1975 to 2007 (Figure 4.9). The analysis of 9856 fires in Trás-os-Montes in the period 1975–2007 resulted in average values for SHAPE = 1.96, indicating a general departure from simple shapes (Figure 4.9).

However, the analysis of the relationship between the perimeter and area in the graphic represented in Figure 4.9 using a log–log scale is more interesting (Figure 4.10).

According to the hypothesis that the shape would remain simple and similar in small and big fires, the exponent of the power equation should be 0.5, as in a circle. As the value of the power estimated (0.626) is higher than the reference value (0.5) the conclusion is that, in effect, shapes of areas burned tend to be more complex as size increases. This could also be observed if we plot the shape indices against size.

This analysis allows for the interpretation of this pattern by studying the underlying processes. In this case, the study of the importance of the various factors in determining the complexity of the shape could be a further step. Another application of this

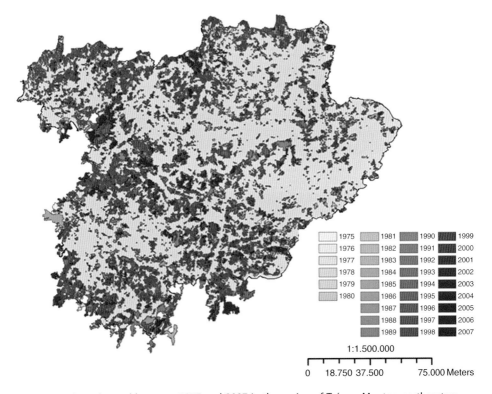

Figure 4.9 Areas burned between 1975 and 2007 in the region of Trás-os-Montes, northeastern Portugal. *Sources:* J.M.C. Pereira and the Portuguese Forest Services.

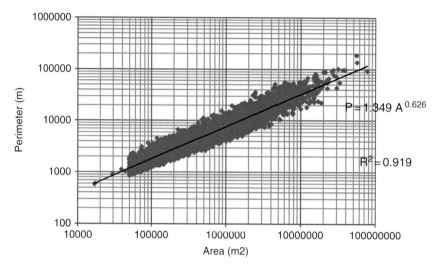

Figure 4.10 Log–log plot (base 10) showing the relationship between the perimeter and area of fires that burned in the Trás-os-Montes region of northeastern Portugal between 1975 and 2007.

relationship would be the estimation of fire perimeter from burned area. This result could be important to fire operation decisions regarding fireline mop-up efforts.

4.6 Patch Interior and Edge

For biodiversity studies it is often important to differentiate between areas in the interior of the patch (core areas) and areas of the patch that are close to the perimeter (edge areas), which can be transition zones to other patches (Figure 4.11).

This concept allows for the use of other metrics to describe shape. In this case, the area of the patch is subdivided into two categories, the area that is within a certain distance from the patch perimeter (EDGE) and the remaining area in the interior (CORE). Obviously, the total area of the patch (A) is the sum:

$$A = EDGE + CORE$$

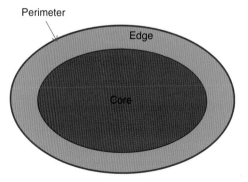

Figure 4.11 The two areas defined within the perimeter of a patch: the CORE (interior) area and the EDGE (transitional) area.

Several indices have been proposed using these measurements to characterize the shape of a patch. One simple index is the ratio between the interior (core) and edge area, as mentioned in a previous example[25]. This interior-to-edge ratio (IEDGERATIO) is simply

$$IEDGERATIO = CORE/EDGE$$

As patches get larger, the ratio of the interior-to-edge area tends to increase as long as the patch shape remains more or less constant, meaning that large patches are mostly interior habitats and small patches, that is those with dimensions less than 2 times the edge width, are all edge habitats. The patch size where the interior area appears represents a threshold above which the area for edge species increases at a much lower rate than the area for interior species. Below this threshold species density (species number per unit area) is generally higher, in spite of containing fewer and, in general, no uncommon species[26].

More often, a core area index (CAI) can then be used as a measure of shape, and is defined as the ratio (or percentage) of interior (core) area to the total area; it can be calculated as a percentage by[27]

$$CAI\,(\%) = 100\,CORE/A = 100(A - EDGE)/A$$

CAI approaches 100% when the patch, because of size, shape, and edge width, contains mostly core area. This would be most common in landscapes with large patches, simple patch shapes approaching squares or circles, and narrow edge widths.

It should be noted that the interpretation of this index is dependent on the definition of the distance from the perimeter that is considered as the edge. That definition has to be made by the user according to the organism, the process, or the objective of the analysis.

Numerous cases of adjacency relationships related to species natural history and general ecology have been described in the literature. For example, Cadenasso and Pickett[28] found distinct relationships between the distance grazing animals would cross for herbivory purposes. The distance varied with the herbivore and the type of vegetation adjacency. A meadow vole (*Microtus pennsylvanicus*) herbivory generally occurred within 75 m of the forest–field edge, whereas a white-tailed deer (*Odocoileus virginicus*) herbivory extended more than 100 m into the field.

Similar examples can be found in studies of plant distributions. For example, Kemp and others[29] found relationships between conifer regeneration and distance to a surviving seed source following a wildfire. Most of the regeneration of the four dominant species occurred within 95 m of a surviving seed-source tree (Figure 4.12).

For studies on edges it is important to note that, when using a raster representation of the patch, the grain or cell size is important because the edge width necessarily has to be a multiple of cell size. Commonly, the total area of the patch (*A*) is measured by the total number of cells in the patch, EDGE is the number of edge cells, and CORE is the number of cells in the interior of the patch. An interior cell is defined as one whose boundary is greater than a predefined distance (edge width) from the polygon edge.

In our example (Figure 4.7) using the edge width as one cell and using the number of cells as units, we have a high percentage of edge cells in the total area ($A = 16$) of the patch: EDGE = 11, CORE = 5. Therefore:

$$CAI = 100(5/16) = 100[(16 - 11)/16] = 31.25\%$$

showing that the higher percentage of the area (68.75%) is in the edge (transition) zone.

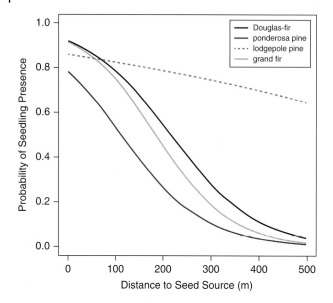

Figure 4.12 Relationship between distance to the seed source and seedling presence for the four conifer species studied in the US Rocky Mountains. *Source:* Kemp, K.B., Higuera, P.E., and Morgan, P. (2015) Fire legacies impact conifer regeneration across environmental gradients in the US northern Rockies. *Landscape Ecology*, 30, 619–636.

Finally, in some special patch configurations, it is possible and informative to characterize shape by the ratio between length (*L*) and width (*W*) of the patch (LWRATIO). This index is often not used since it is difficult to define and to interpret it when shapes are complex. An individual patch may vary in shape from being very simple and compact (close to a square), with LWRATIO approximately equal to 1, to shapes closer to linear elements with values of LWRATIO greater than 10. Some examples in Figure 4.13 illustrate the use of the different indices to characterize the different complexity of patches with the same area (*A* = 16 cells).

The results of the different indices of shape presented before are shown in Table 4.4.

From the results above it is clear that increasing the length/width ratios or the complexity of the shape results in an increased edge habitat and a decreased interior habitat. These have important consequences on the species using those patches.

When calculating indices for various patches of the same class in the landscape, statistics such as means and coefficients of variation are often used. For patch size it is common to compute a mean patch area (MPA) and a patch area coefficient of variation (ACV in percent) as

$$\text{MPA} = \sum a_i/n \quad \text{and} \quad \text{ACV}\,(\%) = 100\left[\sum (a_i - \text{MPA})^2/n\right]^{1/2}/\text{MPA}$$

where a_i is the area of the *i*th patch and *n* is the number of patches considered in the landscape. The coefficient of variation is a dimensionless measure of variability of patch size.

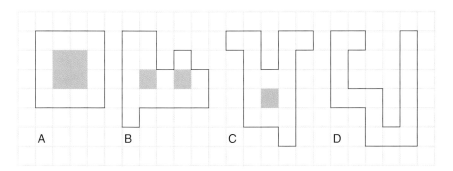

Figure 4.13 Examples of four patches (A, B, C, and D) with the same area (16 cells) and different perimeters (16, 19, 26, and 34 in cell side units from left to right). The core areas (expressed in the number of cells and represented in gray) are 4, 2, 1, and 0.

Table 4.4 The values of the different shape indices calculated for the patches shown in Figure 4.13.

Patch	SHAPE	SHAPE2	CORE	EDGE	IEDGERATIO	CAI (%)	LWRATIO
A	1.00	1.00	4	12	0.333	25.0	4:4 = 1.0
B	1.19	1.17	2	14	0.143	12.5	5:5 = 1.0
C	1.63	1.49	1	15	0.067	6.7	6:5 = 1.2
D	2.13	1.75	0	16	0.000	0.0	6:5 = 1.2

Similar equations can be used for core areas ($CORE_i$) or edge areas ($EDGE_i$) instead of the total patch area (a_i) to compute the mean core size (MCORE) and mean edge (MEDGE) areas and their coefficients of variation (CORECV and EDGECV).

When shape indices are calculated for groups of patches (as for patches of the same class in the landscape), a simple mean core area index (MCAI) or a simple mean shape index (MSI) is often used. However, since larger patches often play a more important role in the function of landscapes, the shape index may be weighted by the area of the patch involved and referred to as the area-weighted mean shape index (AWMSI). The equations are as follows:

$$MSI = \sum(SHAPE_i/n) \quad \text{and} \quad AWMSI = \sum(a_i \times SHAPE_i)/\sum a_i$$

We can illustrate the calculation of indices for a group of patches of the same class in the landscape represented in Figure 4.14.

Table 4.5 summarizes much of the information required for a group of patches of the same class. From this analysis it is important to stress the significance of patch size (as measured by MPA) and its variation (measured by ACV). In fact, patch sizes are variable with areas between 3 and 13. However, from a perspective of an interior species, it is important to know that only 11% (CAI) of the patches are available habitat. It is more of a consequence of the small size of the patches as compared with the width of the edge than it is a result of the shape that is not very complex. This is shown by both MSI and AWMSI, which showed similar low results, indicating relatively simple shapes.

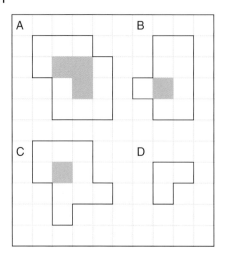

Figure 4.14 Raster representation of a group of four patches of the same class with different sizes and shapes. Core areas are represented in gray.

Table 4.5 Computations for both patch and class metrics from the landscape represented in Figure 4.14.

Patch (i)	Area (A_i)	Core ($CORE_i$)	Core area index $CAI_i(\%)$	Perimeter (P_i)	Shape index ($SHAPE_i$)	Area weighted mean shape index ($A_i \times SHAPE_i$)
A	13	3	0.23	16	1.11	14.42
B	9	1	0.11	14	1.17	10.50
C	10	1	0.10	16	1.26	12.65
D	3	0	0.00	8	1.15	3.46
Sum	35	5	0.44	54	4.69	41.03
Mean	MPA = 35/4 = 8.75	MCORE = 5/4 = 1.25	MCAI = 0.44/4 = 0.11		MSI = 4.69/4 = 1.17	AWMSI = 41.03/35 = 1.17
Coefficient of variation (%)	ACV = 41.5	CORECV = 87.2				

Finally, last but not least, from perspectives of both organisms and processes, the most important measure of significance of any group of patches of class *j* in the landscape is given by its proportion (p_j):

$$p_j = \sum a_{ij}/TA$$

where TA is the total area of the landscape. In our example, $\sum a_{ij} = 35$ and TA = 100. Therefore $p_j = 0.35$.

We will explore the importance of the proportion of the landscape occupied by a patch class in the next section.

4.7   Interaction Between Patches and the Theory of Island Biogeography

The interaction of patches of the same class can be of great importance for both organisms and ecological processes. The implications of these interactions on species diversity were first introduced in 1967 by MacArthur (Figure 4.15) and Wilson[30] when they proposed the theory of island biogeography, where species richness on each island was viewed as a dynamic equilibrium maintained by continued immigration and local extinction of species. Many subsequent studies[31,32,33,34] have also illustrated the importance of immigration to offset the effects of local extinction of mainland populations.

MacArthur[35] further elaborated on the probabilities of extinction by demographic stochasticity and found that, theoretically for a colony founded by one pair, the expected growth rate (dN/dt) and the time to local extinction (T_e) are approximated by

$$dN/dt = (\varepsilon - \mu)N(1 - N/K) \quad \text{and} \quad T_e = (\varepsilon/\mu)^K/(K\lambda)$$

where:

ε is the per capita birth rate,
μ is the per capita mortality rate, and
K is the carrying capacity (the maximum possible colony size at the island).

The survival of a population of a certain species on an island (or patch) would therefore depend on birth (ε) and mortality (μ) rates, and on the maximum carrying capacity of the island for that species (K). The effect on any given species of other species entering the island can be modeled as these new species are often competitors and predators,

Figure 4.15 Robert MacArthur (1930–1972), a North American scientist who made very important contributions in ecology including biogeography. *Source:* https://en.wikipedia.org/wiki/Robert_MacArthur.

resulting in decreased birth rates and increasing mortality, and therefore reducing (ε/μ) and the expected time to local extinction (T_e).

The effect of the maximum possible population size (or the carrying capacity of the island for that species) K is also of extreme importance since a small reduction of the exponent K would drastically reduce the expected survival time of the species. As survival time and probability of extinction are inversely related, MacArthur[36] concluded that the rate of species extinction on an island as new species arrive would be higher on small islands as they are likely to yield lower values of K than on large islands (with higher values of K). The size of the island (or the patch) has therefore a great effect on the rate of local extinctions.

The immigration rate of successful species on to an island (those species that arrive and are able to establish a population) was also considered in general to decrease according to the number of species already existing on the island, as competition between species increases and the availability of niches decreases. However, the immigration rate is also related to the difficulty of reaching the island, or its isolation, which MacArthur and Wilson[37] considered to be well measured by the distance from the mainland source. The immigration rate decreases as more species are established and it is obviously higher on islands that are closer to the mainland (lower distances) than on islands away from the mainland sources (higher distances).

Considering this approach, the dynamic equilibrium of species richness on an island, as a function of the balance between local extinction and immigration rates, could be used to see the effect on species richness of the size of the island and of the distance to the continent.

The theory of island biogeography was presented in 1967 and has been a subject of very intensive debates thereafter (Figure 4.16).

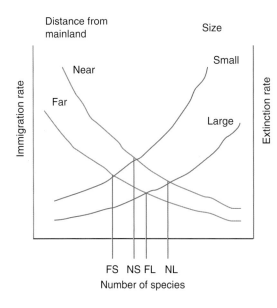

Figure 4.16 A graphical representation of the theory of island biogeography indicating that in general the immigration rate decreases with the number of species already present on the island and the extinction rate increases with the number of species present. The equilibrium number of species is obtained when the immigration rate equals the extinction rate. As islands near mainland (N) have higher immigration rates than islands far from mainland (F) and small islands (S) have higher extinction rates than large islands (L), the number of species on an island that is far from the mainland and small (FS) is much smaller than that of an island that is near the mainland and large (NL). Intermediate values are expected for a small island near the mainland (NS) and large islands far away from the mainland (FL). Adapted from MacArthur, R. and Wilson, E.O. (1967) *The Theory of Island Biogeography*, Princeton University Press, Princeton, NJ.

In spite of the elegance and simplicity of the theory, in reality there are many other factors to take into account. For instance, from various studies it became more and more apparent that the immigration rate depends on the dispersal ability of the taxon and not just the distance from the mainland source[38] (Table 4.6). This becomes obvious if we compare the extreme variability of the maximum oceanic distances crossed, leading to new populations on islands by the main groups of land and freshwater vertebrates[39].

There are also situations where the distance from the mainland factor does not apply. Comparison between the Channel Islands (just off the coast of France) and the Azores shows that the theory applies for nonmarine birds but not for ferns that reproduce using light spores capable of being carried over extremely long distances by the wind. Thus, the immigration rate of ferns is not well measured by the distance from the mainland[40,41].

Of course, island communities are now largely influenced by new immigrants easily dispersed (humans with their accompanying entourage of rats, feral livestock and plants, including weeds). It has been estimated that about 2000 bird species formerly found on tropical Pacific Islands have been driven to extinction by humans and their associated species. This represents a 20% reduction in the world bird species[42].

The theory of island biogeography became dominant in conservation biology and was applied to both actual islands and islands of habitat (patches) in a terrestrial context[43] (Figure 4.17). This was already recognized by MacArthur when he indicated that some

Table 4.6 Widest oceanic gaps crossed during geologic time by various groups of animals.

Animal group	Width of oceanic gap (km)
Freshwater fishes	5
Large mammals	50
Small carnivores	450
Rodents and land tortoises	900
Lizards	1800
Birds, bats, insects, spiders, and land molluscs	3600

Data from various sources: (1) Gorman, M.L. (1979) *Island Ecology*, Chapman and Hall, A Halsted Press Book, John Wiley & Sons, Ltd, London and New York; (2) Menard, H.W. (1986) *Islands*, Scientific American Books, New York; and (3) Huggett, R.J. (2004) *Fundamentals of Biogeography*, 2nd edn, Routledge, Taylor and Francis Group, London and New York.

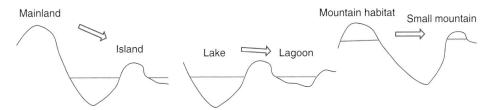

Figure 4.17 Some examples of interactions between habitat patches that are geographically separated and where the theory of island biogeography may apply: from mainland to true islands, from lakes to lagoons, or from large patches of mountain habitats to smaller patches.

mainland habitats were obviously islands, including examples of spruce bogs in deciduous forests, alpine mountaintops, whole mountains separated by deserts, farmers´ woodlots surrounded by agricultural fields, patches of second growth on old growth forests,
or recent fire burns[44].

4.8 Interaction Between Patches and Populations: The Concept of Metapopulation

More recently, and following the early works of Levins[45], the concept of metapopulation
emerged as a valuable approach to deal with small local populations migrating between a
large number of discrete habitat patches, making re-establishment possible following a
local extinction (Figure 4.18). In the simplest Levins´ model, habitat patches are scored
only as occupied or not, and the model therefore applies best to situations where local
populations quickly reach the local "carrying capacity" or where the colonization rate
is low.

The variation of the proportion of occupied patches (p) with time (t) is given by

$$dp/dt = (\sigma)(p)(1-p) - (\phi)(p)$$

where σ and ϕ are the colonization and extinction rate parameters applying to the proportion of occupied patches as sources of colonists (p) and the proportion of empty
patches as targets of colonization ($1 - \sigma\phi$).

The equilibrium value of p (p_{eq}) is given by

$$p_{eq} = 1 - \phi/\sigma$$

showing that the metapopulation can only persist if $\phi/\sigma < 1$, that is, if recolonization can
compensate extinction and the result is a shifting mosaic of occupied and unoccupied

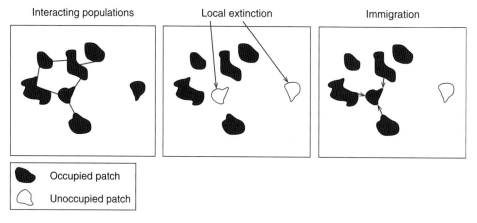

Figure 4.18 Representation of the Levins model of metapopulation dynamics showing patches with
interacting populations and one isolated patch. When local extinction occurs in small patches
immigration from neighbor patches can reoccupy the patch while in isolated small patches the
extinction may last longer.

patches. The Levins model predicts that the risk of metapopulation extinction decreases with increasing p_{eq}, that is, with high colonization rates (σ) from near populations and low extinction rates (ϕ) from increasing patch areas[46].

In summary, it is believed that colonization rates on islands can sometimes be estimated by the distances from the mainland, as in the classical island biogeography theory but, in the case of metapopulations without a mainland, they are a function of the distance from that island (or patch) to its occupied neighbors and their area.

On the other hand, the probability that a given population goes locally extinct is often assessed to be dependent on patch area since it is assumed that there is a linear relationship between patch size and maximum local population size (K).

Finally, for each species, not only dispersal distances but also factors such as birth and death rates, competition ability, and carrying capacity are important to take into account when understanding their distribution and movement between neighboring patches.

A good example highlighting the importance of interaction between patches is provided by a study in Finland on the persistence of a metapopulation of the Glanville fritillary butterfly *(Melitaea cinxia)* (Figure 4.19). In a study by Hanski and others[47] of a network of 1502 dry meadows (suitable habitat patches for the butterfly) was found on 536 of them. They reached the conclusion that the long-term persistence of the butterfly metapopulation in the area was due to the balance between local extinction and recolonizations, since local populations were well defined in meadow patches and too small to be safe from local extinction. They also considered that simultaneous extinction of all local populations was unlikely and that the meadows were not too isolated to prevent recolonization (Table 4.7).

Figure 4.19 The Glanville fritillary butterfly *(Melitaea cinxia)* is an endangered butterfly species, surviving in Finland only within an area of 50 km × 70 km in the Åland islands between Finland and Sweden. *Source: "Melitaea cinxia"* by Niclas Fritzén and Robinson, R. (2006): Genes affect population growth, but the environment determines how. *PLoS Biol,* 4/5/2006: e150 doi:10.1371/journal. pbio.0040150, https://commons.wikimedia.org/wiki/File:Melitaea_cinxia.png#/media/File: Melitaea_cinxia.png.

Table 4.7 Comparison of distances moved by migrating butterflies and between nearest-neighbor suitable habitat patches (dry meadows).

	Distance moved by migrating butterflies (meters)	Distance between suitable habitat patches (meters)
Mean	590	240
Median	330	128
Maximum	3050	3870

Adapted from Hanski, I., Pakkala, T., Kuussaari, M., and Lei, G. (1995) Metapopulation persistence of an endangered butterfly in a fragmented environment. *Oikos*, 72, 21–28.

Table 4.8 Proportion (*p*) of occupied meadows in each 2 km x 2 km square of the geographic range of the butterfly *Melitaea cinxia* in Finland, showing the effect of patch size and density.

Effect of patch size		Effect of patch density	
Mean patch size (ha)	Occupancy (proportion *p*)	Patch density (number of patches per 4 km^2)	Occupancy (proportion *p*)
<0.01	0.24	1	0.21
0.01–0.10	0.24	2–3	0.32
0.10–1.00	0.40	4–7	0.25
>1.00	0.56	>7	0.41

Adapted from Hanski, I., Pakkala, T., Kuussaari, M., and Lei, G. (1995) Metapopulation persistence of an endangered butterfly in a fragmented environment. *Oikos*, 72, 2128.

The conclusion of the study was that the proportion of suitable meadows occupied increases with increasing mean patch size and increasing patch interaction as measured by patch density (Table 4.8).

From the above considerations it is apparent that patch size and distance between patches (as indicated by patch density) are the main characteristics governing the degree of interaction between the local populations.

Many other examples showing the relationships between metapopulation dynamics and persistence and distribution of habitat patches are available in the literature both in conservation biology and landscape ecology journals[48,49].

4.9 Estimating the Interaction Between Patches by the Distance and Size of Neighbors

In previous chapters we demonstrated how a simple measure such as patch density (λ_p), the number of patches per unit area, provides important information about the interaction between patches. Patch density (λ_p) is probably the most commonly used index as a measure for a group of patches of the same class in the landscape. However, patch

Density does not take into account the spatial distribution of patches. The observed mean nearest neighbor distance (OMD), computed as the average of the observed distances of patches of one class to the nearest patch of the same class, is also a meaningful metric. It is generally calculated from the minimum edge-to-edge distance[50]. In this case the pattern of the distribution is not easy to assess as this distance depends not only on density and spatial distribution of patches but also on their size and shape. However, if the patches are small and the centers of the patches are used, the expected mean distance between nearest neighbors (EMD) can be predicted, for a random distribution, from patch density, as already seen in the chapter on the distribution of points. The expected mean distance between nearest neighbors (EMD) would be calculated from patch density (λ_p):

$$\text{EMD} = (1/2)/(\lambda_p)^{0.5}$$

and the index (PATTERN3) is computed as

$$\text{PATTERN3} = \text{EMD}/\text{OMD}$$

This ratio PATTERN3 can be seen, as before, as an index of spatial pattern, ranging from a minimum value of 0.465 for the maximum regularity to infinity for a completely clustered distribution where all points coincide, going through the value of 1.0 for a completely random distribution of patches.

For two patches, the simplest way to indicate their interaction is by measuring their distance, but this fails to take into consideration the areas of the interacting patches.

The same need to consider simultaneously distance and size was in the origin of the simple equation of the proximity index (PROX1) proposed by Gustafson and Parker[51] to distinguish isolated patches from those that are part of a complex of patches. The index (PROX1$_i$) was calculated for each patch using the area (a_i) and distance (d_{ij}) from that patch to the neighbor patches (j) with area a_j of the same class under a specified search radius:

$$\text{PROX1}_i = \sum (a_j/d_{ij})$$

Following these approaches based on areas and distances, other metrics, which also may be meaningful for organisms and many ecological processes, can be computed for an individual patch taking into account the degree of interaction of that patch with all other patches of the same class within a certain search radius. This search radius should be selected by the user as determined by the species (e.g. home range, speed of movement) or the process (e.g. transport of embers in a fire) and has been useful in describing the flow of birds across a landscape or seeds moving between woodlots[52].

Using the similarity of the gravitational force between two bodies, directly proportional to product of their masses and inversely proportional to the square of their distance (Figure 4.20), a simple way to evaluate the strength of the interaction between two patches (I_{ij}) taking into account their areas (a_i and a_j) and their distance (d_{ij}) is through a gravity model measured by

$$I_{i,j} = (a_i \times a_j)/d_{ij}^2$$

A second proximity index for any patch i (PROX2$_i$) in the landscape can thus be defined[53] as

$$\text{PROX2}_i = \sum (a_j/d_{ij}^2)$$

Figure 4.20 The gravitational force between two bodies (planets in this case) depends on their masses and distance. *Source:* Google Earth.

This index considers both the size (a_j) and the distances (d_{ij}) between the focal patch i and all other patches having edges within a sampled search radius of the focal patch.

For a group of patches of the same class in the landscape it is useful to define a mean proximity index (MPROX2)[54]. The value of this index increases with lower isolation of patches and less fragmentation of the class (Figure 4.21). We can compute the mean proximity indices for the n patches of the same class in the landscape:

$$\text{MPROX1} = \sum \text{PROX1}_i/n$$
$$\text{MPROX2} = \sum \text{PROX2}_i/n$$

with a value of 0 if all patches have no neighbors of the same class within the specified search radius, increasing as patches become less isolated and patch class becomes less fragmented. It should only be used as a comparative index.

Following the procedures outlined previously, simple statistics can be computed for the patch class including patch size, perimeter, and shape. However, we are now interested in additional interpatch characteristics that can be calculated. If we consider each square pixel to represent 1 hectare, we have

Patch density (λ) = number of patches (n)/total area $(\text{TA}) = 10/100\,\text{ha} = 0.1\,\text{ha}^{-1}$

Then, for each patch, we can determine the distance to its nearest neighbor and a proximity index. For this latter case we need to define a search radius or a maximum edge-to-edge distance (center-to-center distances may also be used). The search radius selected should be in relation to the organisms, or process, of interest.

Figure 4.21 Example of a group of 10 patches of the same class in a 10 × 10 raster landscape.

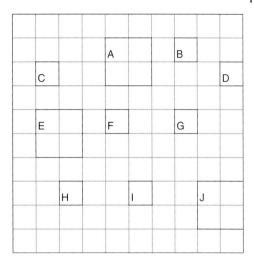

In our example (Figure 4.21) we define the search radius as a distance of 150 m between patch edges. We can illustrate these procedures for patch A. The calculations and results are shown in Table 4.9. For this patch we can observe a minimum (nearest-neighbor) edge-to-edge distance of 100 m.

If we name the patches from A to J (left to right, top to bottom) and use similar procedures for all other patches, we can compute individual proximity indices ($PROX2_i$) for all patches and a mean proximity index (MPROX2) for the whole class (Table 4.10).

We now have the mean patch area (MPA) as 19 000 m^2 (1.9 ha), patch density (λ_p) calculated as 0.1 ha^{-1} (10 patches/100 ha = 10/10^6 m^2), the observed mean nearest-neighbor distance (OMD) as 124.1 m (edge-to-edge) or 265.8 m (center-to-center), and a mean proximity index (MPROX2) as 3.45.

In order to detect the pattern of the spatial distribution of the patches, we can now compute EMD (the expected mean distance between centers of nearest neighbors under a random distribution) as

$$EMD = (1/2)/\lambda_p^{0.5} = (1/2)/(10/10^6 m^2)^{0.5} = 158.1\ m$$

and PATTERN3 (a spatial pattern index) as

$$PATTERN3 = EMD/OMD\ (center-to-center) = 158.1\ m/265.8\ m = 0.59$$

Table 4.9 Calculations for the proximity index of patch A ($PROX2_A$) using the area and distance of its four neighbors with edges closer than 150 m.

Neighbor	Area (m^2)	Distance (m)	Area/ Distance^2
Patch B	10 000	100	1.0
Patch G	10 000	141	0.5
Patch F	10 000	100	1.0
Patch E	40 000	141	2.0
	Proximity index ($PROX2_A$) = 4.5		

Table 4.10 Computations of the interaction indices for all patches and the whole class.

Patch (*i*)	Area (m²)	Nearest neighbor distances (m) Center-to-center	Nearest neighbor distances (m) Edge-to-edge	Proximity Index (PROX2$_j$)
A	40 000	255	100	4.5
B	10 000	224	100	5.0
C	10 000	255	100	4.0
D	10 000	224	100	1.5
E	40 000	255	100	5.0
F	10 000	255	100	8.0
G	10 000	283	141	2.5
H	10 000	255	100	4.0
I	10 000	300	200	0.0
J	40 000	354	200	0.0
Average	19 000	265.8	124.1	MPROX2 = 3.45

showing a strong tendency for the center points of the patches to be regularly distributed (PATTERN3 < 1). These computations illustrate the most commonly used metrics for interactions between patches of the same class.

Several applications of these indices can be found in the literature. In a review of 24 published studies on invertebrates, reptiles, and amphibians metapopulations from around the world are given. Prugh[55] found that the area and the distance to the nearest occupied habitat provided useful data that were easily calculated and easily compared across species and between species. Patch area and within-class patch distance provide an insight into species conservation and are in general agreement with the theory of island biogeography. More complicated measures of patch distance may be more useful when habitat modeling is attempted.

4.10 An Example of the Use of the Gravity Model

The gravity model expresses mathematically the First Law of Geography (generally attributed to Waldo Tobler), that "everything is related to everything else, but near things are more related than distant things" by a general equation:

$$I_{ij} = Ga_i a_j / d_{ij}^k$$

where:

I_{ij} is a measure of the interaction between objects *i* and *j*,
a_i and a_j are the sizes of the two objects,
d_{ij} is the distance between *i* and *j*,
G is a proportionality gravity coefficient, and
k is the distance exponent.

If objects are bodies in space at a distance d, their sizes expressed by their masses, and their interaction measured by the force of attraction, the gravity model can be used with G as the gravitational constant and k the distance exponent as 2. However, the use of the gravity model to estimate the interaction between objects has been used in various research fields where the coefficients G and k are unknown and therefore have to be estimated.

A good illustration is provided by the transportation sector. The importance of considering both size and distance in human flows was in the origin of the hypothesis developed by George Zipf[56], suggesting that migration between two cities i and j was proportional to the product of their populations and inversely proportional to their distance.

In this case the objects can be cities; the size of cities can be measured as their populations and the interactions between patches can be seen as flows or movements of people or goods between the different cities. This was the approach used in the analysis of the interactions between regions and cities in China[57]. The interactions between the 29 regions (I_{ij}) were measured by the quantity of goods carried by railway trains, the city sizes were quantified by the urban population in the region (a_i), and the distance was measured by the railway distances between the capital cities (d_{ij}) (Figure 4.22).

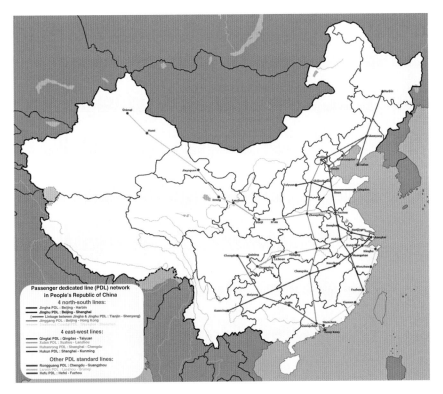

Figure 4.22 The railway network of China. *Source:* https://upload.wikimedia.org/wikipedia/commons/9/97/4%2B4_PDL_network_in_China(English_version).png.

By fitting the theoretical equation to real data it was possible to estimate the gravity model as

$$I_{i,j} = Ga_ia_j/d_{ij}{}^{2.106}$$

showing that the distance exponent was close to 2, as expected from the indices of proximity used.

Other authors have been using similar approaches to estimate the value of the distance exponent, which can have different values, such as the value of 2.5 obtained for Paris and its surroundings[58]. This example shows that the gravity model approach can have different applications, as in the transport sector, where it is possible to use this model to simulate and understand the possible consequences of increases in population on the traffic flows.

In other areas, as in the design of conservation reserves, allowance is made for the simulation of alternative options and their consequences in the movement of the various organisms (typically animals) through the landscape. Analyses made at appropriate scales can therefore be used to test hypotheses concerning alternatives of reserve design for specific species, such as the wild turkey (*Meleagris gallopavo*)[59] or the northern spotted owl (*Strix occidentalis caurina*)[60] (Figure 4.23).

Figure 4.23 The northern spotted owl (*Strix occidentalis caurina*), a threatened species in North America. *Source:* Photo by Paul Bannick.

Key Points

- Patches are defined as relatively homogeneous areas that differ from their surroundings. The description and classification of patches should focus on characteristics of relevant ecological implications for the organism or process of interest, for example vegetation type, soil type, land use category, or habitat characteristics.
- Size is an important characteristic of a patch because the number of species found in a patch is dependent on its size. Larger patches have more "interior habitat", which allows for both species that prefer the interior and those that prefer the edges to occupy the patch.
- Species in smaller habitat patches have been associated with a higher probability of local extinction because of demographic, genetic, and environmental stochasticity and sensitivity to natural disturbances.
- In addition to size, a patch is also characterized by its perimeter and shape. Patch shape has been shown to influence patch microclimate; movement of water, energy, and materials; the spread of disturbances; the recruitment of plants; or foraging strategies and migration of animals.
- Patch shape is calculated by comparing the measured patch perimeter to the simplest perimeter of a patch with the same area, which is a circle for vector data and a square for raster data, or another measure of the perimeter-to-area ratio.
- For studies of diversity it is often important to differentiate between areas in the interior of a patch (core areas) and areas closer to the perimeter (edge areas). The area defined as the core area depends on the organism in question, that is, edge effects vary by species and system.
- Different approaches and metrics are used to quantify area, perimeter, and shape of patches depending on whether they are represented by vector or raster data.
- MacArthur and Wilson's island biogeography theory suggests that the species richness on an island (or patch) depends on the size of the island (patch) and the distance to the mainland (nearest source population). Immigration rates depend on the distance to and size of the source population and extinction rates depend on available resources (often represented by the size of the island).
- Levins' metapopulation theory shows that when local extinction occurs in small patches immigration from neighbor patches can reoccupy the patch while in isolated small patches the extinction may last longer. Again, the size of patches and distance to neighboring patches is important.

Endnotes

1 Forman, R.T.T. (1995) *Land Mosaics. The Ecology of Landscapes and Regions*, Cambridge University Press, Cambridge, MA.
2 Hobbs, R.J. (1993) Effects of landscape fragmentation on ecosystem processes in the western Australian wheatbelt. *Biological Conservation*, 64, 193–201.
3 Rosenzweig, M.L. (1999) *Species Diversity in Space and Time*, Cambridge University Press, Cambridge, MA.
4 Watson H.C. (1847–1859) *Cybele Britannica: Or British Plants and Their Geographical Relations*, Longman, London.

5 Williams, C.B. (1964) *Patterns in the Balance of Nature*, Academic Press. London.

6 Greig-Smith, P. (1983) *Quantitative Plant Ecology*, 3rd edn, University of California Press, Berkeley, CA.

7 MacArthur, R. and Wilson, E.O. (1967) *The Theory of Island Biogeography*, Princeton University Press, Princeton, NJ.

8 Rosenzweig, M.L. (1999) *Species Diversity in Space and Time*, Cambridge University Press, Cambridge, MA.

9 Watson H.C. (1847–1859) *Cybele Britannica: Or British Plants and Their Geographical Relations*, Longman, London.

10 Harestad, A.S. and Bunnell, F.L. (1979) Home range and body weight: A reevaluation. *Ecology*, 60, 389–402.

11 MacArthur, R. (1972) *Geographical Ecology: Patterns in the Distribution of Species*, Princeton University Press, Princeton, NJ.

12 Shaffer, M.L. (1981) Minimum viable population size for species conservation. *BioScience*, 31, 131–134.

13 Barbault, R. and Sastrapradja, S.D. (1995) *Generation, maintenance and loss of biodiversity, in* Global Biodiversity Assessment UNEP, Cambridge University Press, Cambridge, MA.

14 Seal, U.S. (1985) The realities of preserving species in captivity, in *Animal Extinctions: What Everyone Should Know* (ed. R.J. Hoage), Smithsonian Institution Press, Washington, D.C.

15 Forman, R.T.T. (1995) *Land Mosaics. The Ecology of Landscapes and Regions*, Cambridge University Press, Cambridge, MA.

16 Hardt, R.A. and Forman, R.T.T. (1989) Boundary form effects on woody colonization of reclaimed surface mines. *Ecology*, 70, 1252–1260.

17 Forman, R.T.T. and Godron, M. (1986) *Landscape Ecology*, John Wiley & Sons, Inc., New York, NY.

18 Buechner, M. (1989) Are small scale landscape features important factors for field studies of small mammal dispersal sinks?*Landscape Ecology*, 2, 191–199.

19 McGarigal, K. and Marks, B.J. (1995) *FRAGSTATS: Spatial Pattern Analysis Program for Quantifying Landscape Structure*, USDA Forest Service General Technical Report PNW-GTR-351, Pacific Northwest Research Station, Portland, OR.

20 Ackerman, B.B., Leban, F.A., Samuel M.D., and Garton, E.O. (1990) *User's Manual for Program HOME RANGE*, 2nd edn, Forestry, Wildlife and Range Experiment Station Technical Report 15, University of Idaho, Moscow, ID.

21 Forman, R.T.T., Galli, A.E., and Leck, C.F. (1976) Forest size and avian diversity in New Jersey woodlots with some land use implications. *Oecologia*, 26, 1–8.

22 Forman, R.T.T., Galli, A.E., and Leck, C.F. (1976) Forest size and avian diversity in New Jersey woodlots with some land use implications. *Oecologia*, 26, 1–8.

23 Schoener, T.W. (1968) Sizes of feeding territories among birds. *Ecology*, 49, 123–141.

24 Ackerman, B.B., Leban, F.A., Samuel M.D., and Garton, E.O. (1990) *User's Manual for Program HOME RANGE*, 2nd edn, Forestry, Wildlife and Range Experiment Station Technical Report 15, University of Idaho, Moscow, ID.

25 Forman, R.T.T., Galli, A.E., and Leck, C.F. (1976) Forest size and avian diversity in New Jersey woodlots with some land use implications. *Oecologia*, 26, 1–8.

26 Forman, R.T.T. (1995) *Land Mosaics. The Ecology of Landscapes and Regions*, Cambridge University Press, Cambridge, MA.

27 McGarigal, K. and Marks, B.J. (1995) *FRAGSTATS: Spatial Pattern Analysis Program for Quantifying Landscape Structure*, USDA Forest Service General Technical Report PNW-GTR-351, Pacific Northwest Research Station, Portland, OR.

28 Cadenasso, M.L. and Pickett, S.T.A. (2000) Linking forest edge structure to edge function: mediation of herbivore damage. *Journal of Ecology*, 88, 31–34.

29 Kemp, K.B., Higuera, P.E., and Morgan, P. (2015) Fire legacies impact conifer regeneration across environmental gradients in the US northern Rockies. *Landscape Ecology*, 30, 619–636.

30 MacArthur, R. and Wilson, E.O. (1967) *The Theory of Island Biogeography*, Princeton University Press, Princeton, NJ.

31 Den Boer, P.J. (1981) The survival of populations in a heterogeneous and variable environment. *Oecologia (Berlin)*, 50, 39–53.

32 Middleton, J. and Merriam, G. (1981) Woodland mice in a farmland mosaic. *Journal of Applied Ecology*, 18, 703–710.

33 Henderson, M.T., Merriam, G., and Wegner, J. (1985) Patchy environments and species survival: chipmunks in an agricultural mosaic. *Biological Conservation*, 31, 95–105.

34 Pimm, S.L. (1991) *The balance of nature? Ecological issues in the conservation of species and communities*, University of Chicago Press, Chicago, IL.

35 MacArthur, R. (1972) *Geographical Ecology: Patterns in the Distribution of Species*, Princeton University Press, Princeton, NJ.

36 MacArthur, R. (1972) *Geographical Ecology: Patterns in the Distribution of Species*, Princeton University Press, Princeton, NJ.

37 MacArthur, R. and Wilson, E.O. (1967) *The Theory of Island Biogeography*, Princeton University Press, Princeton, NJ.

38 Wilcox, B.A. (1978) Supersaturated island faunas: A species-age relationship for lizards on post-Pleistocene land-bridge islands. *Science*, 199, 996–998.

39 Menard, H.W. (1986) *Islands*, Scientific American Books, New York, NY.

40 Williamson, M.H. (1981) *Island Populations*, Oxford University Press, Oxford.

41 Rosenzweig, M.L. (1999) *Species Diversity in Space and Time*, Cambridge University Press, Cambridge, MA.

42 Steadman, D.W. (1995) Prehistoric extinctions of Pacific island birds biodiversity meets zooarcheology. *Science*, 275, 1123–1131.

43 Prugh, L.R. (2009) An evaluation of patch connectivity measures. *Ecological Applications*, 19, 1300–1310.

44 MacArthur, R.H. (1972) *Geographical Ecology. Patterns in the Distributions of Species*, Princeton University Press, Princeton, NJ.

45 Levins, R. (1970) Extinction, in *Some Mathematical Problems in Biology* (ed. M. Gerstenhaber), American Mathematical Society, Providence, RI.

46 Hanski, I. (1997) Metapopulation dynamics: from concepts and observations to predictive models, in *Metapopulation Biology. Ecology, Genetics and Evolution* (eds I. Hanski and M.e. Gilpin), Academic Press, San Diego, CA.

47 Hanski, I., Pakkala, T., Kuussaari, M., and Lei, G. (1995) Metapopulation persistence of an endangered butterfly in a fragmented environment. *Oikos*, 72, 21–28.

48 Quintana-Asencio, P.F. and Menges, E.S. (1995) Inferring metapopulation dynamics from patch-level incidence of Florida scrub plants. *Conservation Biology*, 10, 1210–1219.

49 Lindenmayer, D.B. and Possingham, H.P. (1996) Modelling the inter-relationship between habitat patchiness, dispersal capability and metapopulation persistence of the endangered species, Leadbeater's possum, in south-eastern Australia. *Landscape Ecology*, 11, 79–105.

50 McGarigal, K. and Marks, B.J. (1995) *FRAGSTATS: Spatial Pattern Analysis Program for Quantifying Landscape Structure*, USDA Forest Service General Technical Report PNW-GTR-351, Pacific Northwest Research Station, Portland, OR.

51 Gustafson, E.J. and Parker, G.R. (1992) Relationship between landcover proportion and índices of landscape spatial pattern. *Landscape Ecology*, 7, 101–110.

52 Forman, R.T.T. and Godron, M. (1986) *Landscape Ecology*, John Wiley and Sons, Inc., New York, NY.

53 Whitcomb, R.F., Robbins, C.S., Lynch, J.F., *et al.* (1981) Effects of forest fragmentation on Avifauna of the Eastern Deciduous Forests, in *Forest Island Dynamics in Man-Dominated Landscapes* (eds R.L. Burgess and D.M. Sharpe), Springer-Verlag, New York, NY.

54 Whitcomb, R.F., Robbins, C.S., Lynch, J.F., *et al.* (1981) Effects of forest fragmentation on Avifauna of the Eastern Deciduous Forests, in *Forest Island Dynamics in Man-Dominated Landscapes* (eds R.L. Burgess and D.M. Sharpe), Springer-Verlag, New York, NY.

55 Prugh, L.R. (2009) An evaluation of patch connectivity measures. *Ecological Applications*, 19, 1300–1310.

56 Zipf, G.K. (1949) *Human Behaviour and the Principle of Least Effort*, Addison-Wesley, Reading, MA.

57 Chen, Y. (2017) The distance–decay function of geographical gravity model: power law or exponential law. Retrieved at: https://arxiv.org/ftp/arxiv/papers/1503/1503.02915.pdf.

58 Rybski, D., Ros, A.G.C., and Kropp, J.P. (2013) Distance-weighted city growth. *Physical Review E*, 87, 042114.

59 Gustafson, E.J., Parker, G.R., and Backs, S.E. (1994) Evaluating spatial pattern of wildlife habitat: a case study of the wild turkey (*Meleagris gallopavo*). *American Midland Naturalist*, 131, 24–33.

60 Murphy, D.D. and Noon, B.R. (1992) Integrating scientific methods with habitat conservation planning: reserve design for northern spotted owls. *Ecological Applications*, 2, 3–17.

5

The Vertical Dimension of Landscapes

So far we have been concentrating on the representation of landscapes by two dimensional maps. However, the third dimension (vertical) is also very important in the distribution of species and populations, as well as in many ecological processes.

From the island biogeography theory we concluded that island area and distance to the mainland were two fundamental drivers of species diversity. It is important to recognize, however, that the vertical dimension can also play a role. In fact, it is known from various studies[1] that islands with higher elevations are generally more diverse in species and habitats than islands with equivalent area and distance to the mainland.

5.1 The Importance of Elevation Illustrated for Birds in the Macaronesian Islands

Because islands are excellent laboratories to study biogeography, as MacArthur and previously Darwin have pointed out, an example of Macaronesia (the enchanted islands) may be used to illustrate the issue of elevation as a third dimension (Figure 5.1).

The effect of elevation can be illustrated by the study of Rosário[2] to understand the factors determining diversity and endemism of the birds of Macaronesian's islands, some of which have high elevations (Figure 5.2).

Rosário[3] found that birds of the Macaronesian islands were sensitive to the available habitats and some types of habitat were only available at higher elevations. Thus some important endemic bird species such as the threatened priolo (*Pyrrhula murina*) (Figure 5.3) were only found in those habitats typical of higher elevations that need a special conservation effort.

From the analysis of the data in Table 5.1 it is clear that the number of species of nesting birds on an island (S) was strongly associated with the area of the island (A), as typically expressed in a log–log plot as represented in Figure 5.4. The positive coefficient for ln (A) (+0.178) indicates that, as expected, species richness increases with area. The percentage of variation of bird richness explained by the area of the island alone (78.6%) was very high, showing the fundamental role of area in determining biodiversity.

Applied Landscape Ecology, First Edition. Francisco Castro Rego, Stephen C. Bunting, Eva Kristina Strand and Paulo Godinho-Ferreira.
© 2019 John Wiley & Sons Ltd. Published 2019 by John Wiley & Sons Ltd.

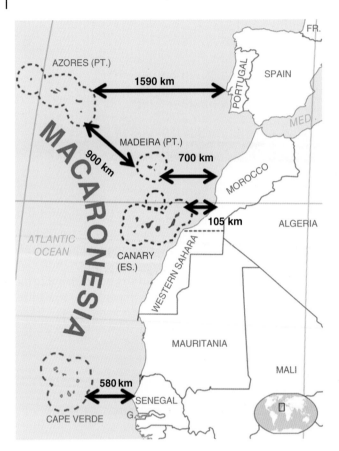

Figure 5.1 The Macaronesian islands including the archipelagos of Azores, Madeira, Canaries and Cape Verde, showing variable distances to the mainland (Iberian Peninsula or Africa). *Source of base map:* Wikipedia, https://www.google.com/search?q=wiki+maps+azores&source=lnms&tbm=isch&sa= X&ved=0ahUKEwjZiJySiuXYAhVMZKwKHZV7AhAQ_AUICygC&biw=1600&bih=767#imgrc= TwMDogt7TNkKAM.

However, we also saw in the theory of island biogeography that the distance to the mainland can play an important role. If we add the effect of distance to mainland (D_m in km) in the equation we would have

$$\ln(S) = 3.325 + 0.171 \ln(A) - 0.152 \ln(D_m) \quad \text{with } R^2 = 0.842$$

showing that there is still some further explanation of the variation of the number of bird species (from 78.6 to 84.2%) due to changes in distance to the mainland. The negative sign in the coefficient for distance (−0.152) shows that, for the same area, the further the island is from the mainland the lower its richness in birds.

This effect of distance from the continent is well known in Macaronesia. In fact it has been known for many years that the species diversity of the archipelago of Azores is lower than that of Madeira due to its greater distance to the mainland.

Figure 5.2 The third (vertical) dimension in the island of Pico, Azores, with 2351 meters elevation. *Source:* https://commons.wikimedia.org/wiki/File:Ilha_do_Pico_vista_da_Faj%C3%A3_Grande,_Calheta,_ilha_de_S%C3%A3o_Jorge,_A%C3%A7ores,_Portugal.jpg

Figure 5.3 The priolo (*Pyrrhula murina*)[4], a threatened endemic bird species of the Azores, has its preferential habitats in the heaths and juniper forests typical of higher elevations. *Source:* Pedro Monteiro, The Portuguese Society for the Study of Birds/Sociedade Portuguesa para o Estudo das Aves (SPEA), http://life-priolo.spea.pt/fotos/ambiente/7889c07c7b9c4a964c9431391cd2215eambiente_priolo.jpg.

What was observed for birds also applies to plant species. In fact, birds assisted colonization of the Azores archipelago through transport of seeds of species with berries as fruit likely played a role in the diversity of plant species and the diversity of floral and other characteristics of populations of endemic species such as the Azores blueberry (*Vaccinium cylindraceum*) (Figure 5.5).

In fact, in distant oceanic islands only a restricted and independent subset of plants and pollinators arrive during early stages of colonization. The effect of distance on the diversity of plants in the Azores archipelago could then be explained by the concept of double

Table 5.1 Data adapted from Rosário[5] relating the number of bird species with the area, the distance to mainland, and the elevation of the islands of Macaronesia.

Archipelago	Island	Species of birds, S (number)	Area of island, A (km²)	Distance to mainland, D_m (km)	Maximum elevation, E (m)
Açores	Corvo	17	17	2190	777
	Flores	22	143	2190	915
	Graciosa	21	62	1915	402
	Terceira	21	382	1795	1021
	S. Jorge	24	246	1860	1053
	Faial	22	173	1950	1043
	Pico	21	446	1890	2351
	S. Miguel	24	759	1610	1105
	Formigas	5	0.2	1590	20
	Sta. Maria	22	97	1620	590
Madeira	Porto Santo	25	40	700	517
	Madeira	39	765	720	1861
	Deserta Grande	17	11	700	479
	Bugio	15	3	700	384
Selvagens	Selvagem Grande	8	2.5	400	136
	Selvagem Pequena	8	0.2	400	49
Canarias	Alegranza	24	11	175	289
	Lanzarote	36	862	134	671
	Fuerteventura	35	1663	105	807
	Gran Canaria	54	1531	210	1950
	Tenerife	53	2055	300	3718
	La Gomera	41	371	350	1487
	La Palma	40	729	440	2423
	Hierro	35	287	400	1512
Cabo Verde	Santo Antão	23	779	840	1979
	S. Vicente	22	227	820	725
	Sta. Luzia	16	35	805	395
	Branco	14	3	795	327
	Raso	15	7	785	164
	S. Nicolau	23	343	735	1304
	Sal	24	216	610	406
	Boavista	30	620	580	387

Table 5.1 (Continued)

Archipelago	Island	Species of birds, S (number)	Area of island, A (km²)	Distance to mainland, D_m (km)	Maximum elevation, E (m)
	Maio	23	269	610	436
	Santiago	32	991	650	1392
	Fogo	25	476	735	2829
	Rombos	15	4	770	95
	Brava	24	64	770	976

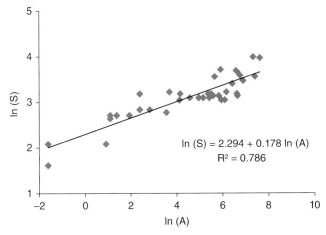

Figure 5.4 Log–log plot (using natural logarithms – ln) of the number of bird species (S) in the Macaronesian islands as related to the size of the islands (A in km²).

insularity proposed by Dias[6], suggesting that seeds have arrived with birds not directly from the nearest continent (about 1600 km) but from the nearest islands (Madeira or other islands not necessarily existing presently).

Finally, we should consider the third dimension and evaluate the possible effects of elevation in the richness of species in the islands. This effect can be expressed by the maximum elevation in the highland (E in meters). Adding this term in the above equation we have

$$\ln(S) = 2.752 + 0.106 \ln(A) - 0.176 \ln(D_m) + 0.159 \ln(E) \quad \text{with } R^2 = 0.873$$

Again, there was also a small but significant contribution of elevation to the explanation of the richness in bird species in the islands, with the percentage of explanation going from 84.2 to 87.3%. The positive coefficient for elevation (+0.159) shows that islands with higher elevations will have higher species richness for similar area and distance to the mainland.

Figure 5.5 The Azores blueberry (*Vaccinium cylindraceum*), an endemic species of the Azores. *Source:* Pereira, M.J. (2008) Reproductive biology of *Vaccinium cylindraceum* (Ericaceae), an endemic species of the Azores archipelago. *Botany*, 86, 359–366.

5.2 Montane Islands

Elevation increases species diversity of islands as well as in mainland mountains. In general, higher elevation results in higher diversity of favorable habitats and thus more options for species having different habitat requirements. McArthur[7] described in 1972 montane islands as special cases because the mountainous areas were more isolated than their lowland counterparts and had greater extinction rates and lower immigration rates, resulting in vegetation that has substantially different characteristics from the surrounding lowlands[8].

In many ways the isolated mountaintops of the American Southwest and Great Basin are subject to isolation processes similar to those occurring in true ocean islands and have been therefore referred to as "sky islands"[9].

However, studies on montane mammals[10] and butterflies[11] have suggested that there is now evidence that it may be appropriate to modify existing general paradigms of the biogeography of montane faunas in the Great Basin as the mountain tops were not as isolated as originally thought because they became isolated progressively as climate was getting warmer. In fact, the cool and wet climate that allowed pine and juniper forests in lower elevations ended only after the last ice age, when the forests started to be restricted to the mountain tops surrounded by a sea of unfavourable dry habitats (Figures 5.6 and 5.7). Similar conclusions were reached with a jumping spider (*Habronattus pugillis*) that lives on oak woodlands restricted to mountain ranges in southern Arizona (Figure 5.8). The remarkable differentiation found among the various sky-island populations was considered to probably be only a few tens of thousands of years old, since the last contact of oak woodlands, although the authors also indicate the possibility that it could be much older, having persisted despite some interpopulation contact.

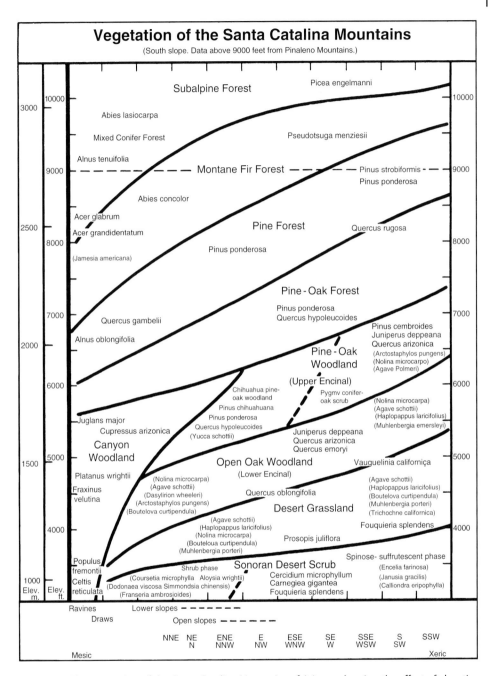

Figure 5.6 The vegetation of the Santa Catalina Mountains of Arizona showing the effect of elevation and aspect[12]. *Source:* Whittaker, R.H. and Niering, W.A. (1965) Vegetation of the Santa Catalina Mountains, Arizona: A gradient analysis of the south slope. *Ecology*, 46, 429–452.

Figure 5.7 Lines of Saguaro cacti (*Carnegiea gigantea*), a typical species of the desert vegetation, with the background of the Santa Catalina Mountains near Tucson, Arizona. *Source:* https://visionarywild. com/wp/wp-content/uploads/2011/05/Dykinga_AZ-010331-0341.jpg.

Many other studies in various parts of the World and with different taxonomic groups have reached similar conclusions about the genetic consequences of the isolation of populations in sky islands. Populations of passerine birds in the Pantepui region[13], poison dart frogs in sky islands in central Peru[14], and starred robins (*Pogonocichla stellata*) in the East African Arc[15] have all been shown to have diverged at either the subspecies or species level[16].

5.3   The Vertical Dimension in Aquatic Systems

The vertical dimension is also an important consideration for habitats and species distributions within rivers, streams, and other water bodies.

Water depth, together with water velocity and sediment size, has been found to be one of the most significant variables explaining distributions of fish in 28 regulated and non-regulated rivers across six physiographic regions in Canada[17].

In regulated rivers the water depth and velocity can vary greatly depending on the discharge at the dam. Studies from the John Day Reservoir on the Columbia River can illustrate this (Figures 5.9, 5.10, and 5.11).

Point data describing river depth can be obtained from echo-sounding or remote sensing with laser technology. The point data can be processed in geographic information systems to create contours or surfaces of the water body floor.

5.4   The Vertical Structure of Vegetation and Species Diversity

Structural vertical heterogeneity of vegetation has been found to increase animal species richness and abundance in many studies. Not surprisingly, the vertical dimension and importance of three-dimensional heterogeneity is also important for flying vertebrates

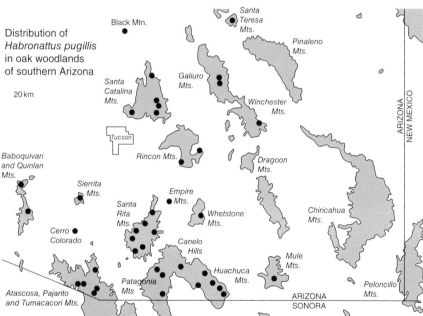

Figure 5.8 The jumping spider (*Habronattus pugilis*) occurring in oak woodland habitats in the mountain ranges of southern Arizona. *Sources:* https://commons.wikimedia.org/wiki/File: Habronattus_pugillis,_adult_male.jpg; Author Thomas Shahan (2012): http://www.flickr.com/photos/opoterser/7708366952/; Maddison, W. and McMahon, M. (2000) Divergence and reticulation among montane populations of a jumping spider (*Habronattus pugillis* Griswold). *Systematic Biology*, 49, 400–421.

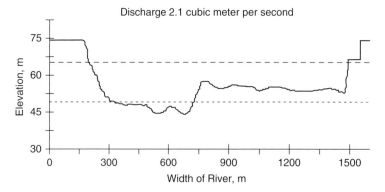

Figure 5.9 The water depth in a cross-section of the Columbia River at two water flow scenarios: natural flow and at a 67 m level of the John Day Reservoir. As the depth changes, the water velocity will also vary greatly from 2.56 m/s in the natural flow scenario to 0.12 m/s in the 67 m level scenario. *Source:* Eva Strand, University of Idaho.

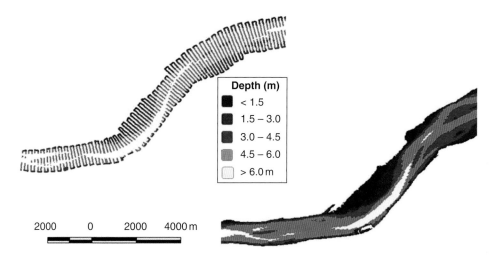

Figure 5.10 A stretch of a few kilometers in the John Day Reservoir of the Columbia River showing a map of the echo-soundings and an interpolated map. *Source:* Eva Strand, University of Idaho.

such as birds and bats. In an extensive review, 20 of 23 avian studies found a higher bird richness and abundance in canopies with high structural variability and complexity[18].

A good example of these relationships is provided by Vogeler[19], who showed that the vertical dimension explains a large portion of the variance in bird richness in two study areas in a mixed conifer forest on Moscow Mountain and Slate Creek in north-central Idaho. In this study, the vertical dimension was characterized by understorey density, canopy density, and height variability while the horizontal dimension was characterized by patch diversity (Figure 5.12). Topography was characterized by slope and elevation and greenness was characterized by the remotely sensed normalized difference vegetation index (NDVI).

Figure 5.11 The John Day River, which flows into the Columbia River, is one of the longest free-flowing rivers of the American West. *Source:* David Jansen Photography, http://www.westernrivers.org/images/project/71/drift-boat-john-day-river.jpg.

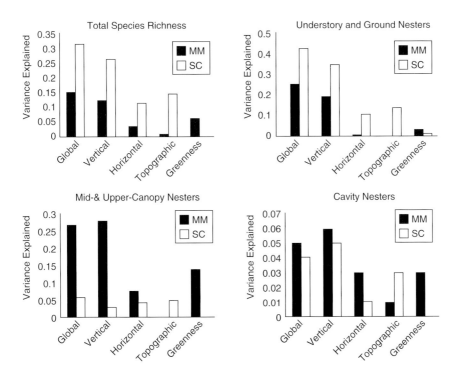

Figure 5.12 Global variance and variance explained by various factors (vertical and horizontal dimensions, topography, and greenness) for different groups of bird species for both Moscow Mountain (MM) and Slate Creek (SC) in north-central Idaho. *Source:* Vogeler, J.C., Hudak, A.T., Vierling, L.A., *et al.* (2014) Terrain and structural influences on local avian species richness in two mixed-conifer forests. *Remote Sensing of Environment*, 147, 13–22.

Figure 5.13 Cross-section view of beech forest from high-resolution airborne LiDAR, Hainich National Park, Germany. *Source:* Courtesy of Shaun Levick, https://bgc-jena.mpg.de/bgp/slideshow/slideshow/slideshow.html.

Recent advances in remote sensing light detection and ranging (LiDAR) technology has made it possible to accurately measure the vertical dimension of vegetation structure and map ecosystems in three dimensions. LiDAR has the ability to quantify canopy cover, canopy height, vertical layering profile, understory vegetation, and topography (Figure 5.13).

These recent technological advances have contributed to a better understanding of the importance of vegetation structure. Many studies have confirmed the importance of the third dimension in creating more habitats and more niches for more species.

In this chapter examples of the importance of the vertical dimension were presented for elevation in islands, for depth in water bodies, and for vertical distribution of biomass in forests. Many other examples could be shown of the relationships between vertical structure and biological diversity and also between vertical structure and many landscape processes.

Key Points

- The vertical dimension has been shown to impact diversity and habitat use. For example, islands with higher elevations are generally more diverse in species and habitats than islands with equivalent area and distance to mainland.
- In aquatic systems, the vertical dimension, that is the depth of the water, is an important characteristic of habitat for aquatic species.
- The vertical dimension in forest canopies has been shown to impact bird richness and abundance. The vertical dimension of forest canopies can be characterized with remote sensing light detection and ranging (LiDAR) technology.

Endnotes

1 Ricklefs, R.E. and Lovette, I.J. (1999) The roles of island area per se and habitat diversity in the species–area relationships of four Lesser Antillean faunal groups. *Journal of Animal Ecology*, 68, 1142–1160.

2 Rosário, L.P. (1996) *Biogeografia Macaronésica: Factores de Diversidade e Endemicidade na Avifauna*. Working Paper Number 49, Curso de Informática da estatística e da Econometria, Instituto Superior de Estastística e Gestão de Informação, Universidade Nova de Lisboa.

3 Rosário, L.P. (1996) *Biogeografia Macaronésica: Factores de Diversidade e Endemicidade na Avifauna*. Working Paper Number 49, Curso de Informática da estatística e da Econometria, Instituto Superior de Estastística e Gestão de Informação, Universidade Nova de Lisboa.

4 http://life-priolo.spea.pt/en/.

5 Rosário, L.P. (1996) *Biogeografia macaronésica: Factores de diversidade e endemicidade na avifauna*. Working Paper Number 49. Curso de Informática da estatística e da Econometria. Instituto Superior de Estastística e Gestão de Informação, Universidade Nova de Lisboa.

6 Dias, E. (1996) *Vegetação Natural dos Açores. Ecologia e Sintaxonomia das Florestas Naturais*, PhD Dissertation, Department of Botany, the University of Azores, Ponta Delgada.

7 MacArthur, R.H. (1972) *Geographical Ecology: Patterns in Distributions of Species*, Harper & Row, New York, NY.

8 Lomolino, M.V. (2001) Elevation gradients of species-density: historical and prospective views. *Global Ecology and Biogeography*, 10, 3–13.

9 Brown, J.H. (1971) Mammals on mountaintops: nonequilibrium insular biogeography. *The American Naturalist*, 105, 467–478.

10 Floyd, C.H., Van Vuren, D.H., and May, B. (2005) Marmots on Great Basin mountaintops: using genetics to test a biogeographic paradigm. *Ecology*, 86, 2145–2153.

11 Fleishman, E., Austin, G.T., and Murphy, D.D. (2001) Biogeography of Great Basin butterflies: revisiting patterns, paradigms, and climate change scenarios. *Biological Journal of the Linnean Society*, 74, 501–515.

12 Whittaker, R.H. and Niering, W.A. (1965) Vegetation of the Santa Catalina Mountains, Arizona: A gradient analysis of the south slope. *Ecology*, 46, 429–452.

13 Perez-Eman, J.L. (2005) Molecular phylogenetics and biogeography of the neotropical redstarts (*Myioborus*; Aves, *Parulinae*). *Molecular Phylogenetic Evolution*, 37, 511–528.

14 Roberts, J.L., Brown, J.L., Shulte, R., *et al.* (2006) Rapid diversification of colouration among populations of a poison frog isolated on sky peninsulas in the Central Cordilleras of Peru. *Journal of Biogeography*, 34, 417–426.

15 Bowie, R.C., Fjeldsa, J., Hackett, S.J., *et al.* (2006) Coalescent models reveal the relative roles of ancestral polymorphism vicariance, and dispersal in shaping phylogeographical structure of an African montane forest robin. *Molecular Phylogenetic Evolution*, 38, 171–188.

16 Department of Biological Science, California State University, Fullerton, Course on Desert Ecology, 2016; http://csufdesertecology.weebly.com/.

17 Senay C., Macnaughton C.J., and Lanthier, G. (2015) Identifying key environmental variables shaping within-river fish distribution patterns. *Aquatic Sciences*, 77, 709–721.

18 Davies, A.B. and Asner, G.P. (2014) Advances in animal ecology from 3D-LiDAR ecosystem mapping. *Trends in Ecology and Evolution*, 29, 681–691.

19 Vogeler J.C., Hudak A.T., Vierling L.A., *et al.* (2014) Terrain and structural influences on local avian species richness in two mixed-conifer forests. *Remote Sensing of Environment*, 147, 13–22.

6

Movements Through Landscapes

One of the most important aspects of the study of the ecology of landscapes is in the understanding of the influence of pattern on the movement of organisms or processes through the landscape. The ecological importance of pattern has long been recognized. For example, foresters have used particular forest harvest techniques to enhance post-harvest tree regeneration. Wildlife managers have realized that particular combinations of habitat adjacent or nearby to one another are required for the survival of some species. Fire managers have known that particular vegetation patterns can either enhance or retard fire movement across the landscape. However, it was not until the advent of landscape ecology as a discipline, and subsequent analysis methods, that we have had comprehensive means to describe and analyze these patterns. A number of approaches used in other disciplines have been implemented to understand movements across landscapes. Network analysis, discussed in Chapter 3, has been applied to the movement of organisms. Other approaches such as percolation and contagion theory are also discussed in the context of movements through landscapes.

6.1 Percolation Theory

In physics, chemistry, and material sciences, the movement and filtering of fluids through porous materials is referred to as percolation. Percolation theory, a branch of physics that investigates the flow of particles or energy through a porous lattice of grid cells[1], was originally applied to porous materials such as soils and ceramics before finding applications in landscape ecology.

In soil science, the downward movement of water within the soil matrix is called percolation and the pore space in soil is the conduit that allows water to percolate (Figure 6.1).

The analogy with soil percolation was found when trying to understand the movement of animals or disturbance in relation to the spatial aggregation of suitable and unsuitable habitat patches[2]. As water percolates through the pores in soil, organisms or matter can flow in the landscapes through favorable habitat patches analogous to pores.

In order to develop a percolation theory we first need to have a model for the landscape. The most common model of a landscape in a percolation theory context uses a square lattice cell grid, with cells classified as suitable or unsuitable for the movement of the

Applied Landscape Ecology, First Edition. Francisco Castro Rego, Stephen C. Bunting, Eva Kristina Strand and Paulo Godinho-Ferreira.
© 2019 John Wiley & Sons Ltd. Published 2019 by John Wiley & Sons Ltd.

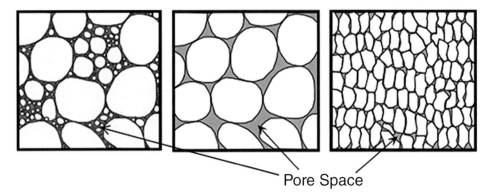

Pore Space

Figure 6.1 Pore space in soils of different textures. Coarse-grained soils (in the middle) usually have a higher porosity than fine-grained soils (right) or mixed-grain soils (left) and thus a higher percolation rate. *Source:* Lectures of Physical Geology EENS 1110 by Prof. Stephen A. Nelson of Tulane University, http://www.tulane.edu/~sanelson/eens1110/groundwater.htm.

organism or entity of interest. In this application it is important to remember that the cells only need to be suitable for movement and may not necessarily constitute a habitat.

A method used to generate landscapes to be analyzed through the percolation theory is by randomly assigning cells as suitable or unsuitable with a statistical independent probability p (Figure 6.2).

These virtual landscapes are built by "neutral" models, that is, the landscape pattern is generated by random processes. These simple random maps can be used as "null hypotheses" to compare with actual landscapes that are generally structured by processes that generate nonrandom patterns. The analysis of these random landscapes allows for the understanding of some general patterns, indicating that in these cases the number, size, and shape of patches of a class are only a function of the proportion of the landscape occupied by that class[3]. The results of the analysis of the maps in Figure 6.2 are displayed in Figure 6.3 and are very similar to those obtained in other studies[4].

However, we are especially interested in the possibility of movement in the landscape and we therefore have to understand how percolation theory can assist in estimating that movement. First, however, we need to define percolation in a square grid lattice.

It is important to distinguish between bond and site percolation. In bond percolation the movement can only be done to four neighbors through the bonds of adjacent cells (edges) while in site percolation the movement can also be done through sites (the vertices of the cells) with each cell having eight neighbors (Figure 6.4). We will be using mostly bond percolation, except when explicitly mentioned otherwise.

The possibility of percolation can then be assessed from the existence of a path between sides of cells of favorable habitat (bond percolation) or by sides and vertices (site percolation) (Figure 6.5).

Percolation theory addressed the issue of landscapes being favorable or not favorable to movement by defining a critical threshold (p_c), where large favorable percolating clusters are formed and long-range connectivity first appears. This percolation threshold is, however, dependent on the method used. While for site percolation the proportion

Figure 6.2 A landscape of 20 × 20 cells randomly "occupied" by a suitable habitat for movement (dark cells) with decreasing probabilities (0.7 at the top left to 0.2 in the bottom right with steps of 0.1). The actual proportions resulting from the random assignment were, from left to right and up to down, 0.6825, 0.5925, 0.465, 0.3525, 0.285, and 0.185.

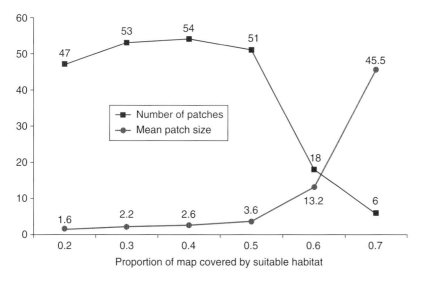

Figure 6.3 Results of the analysis of the six random landscapes shown in Figure 6.2, showing that as the proportion (p) of the suitable habitat increases the number of patches increases initially, but when $p >$ 0.4 the habitat becomes less fragmented and the number of patches begins to decline whereas the mean patch size increases rapidly for $p > 0.5$. Both variables (number of patches and mean patch size) are represented in the same Y-axis.

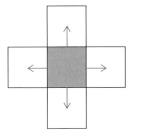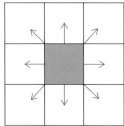

Figure 6.4 The four-neighbor rule (left) used for bond percolation (movement only through the edges of the cell) and the eight-neighbor rule (right) used for site percolation (movement can also be made through the vertices of the cells).

Figure 6.5 The two first random landscapes represented in Figure 6.2 with a proportion of favorable habitats of 0.6825 and 0.5925, respectively. It is apparent in the first landscape that it is possible to move from one side to the other of the landscape through sides of contiguous gray cells (bond percolation is possible). In the second landscape bond percolation is not possible but site percolation exists. In both cases possible paths from left to right of the landscape are shown as white lines.

threshold is defined as $p_c = 0.50$, the threshold for bond percolation was empirically determined for large maps by various authors[5,6] to be $p_c = 0.593$.

The example shown in Figure 6.5 can illustrate this issue. The first two landscapes, with a proportion of favorable habitat above the critical threshold of 0.50, allow for site percolation whereas the only landscape allowing for bond percolation is that where the proportion of favorable habitat is 0.6825, which is above the theoretical threshold of 0.593.

Regardless of the method used it is easy to conclude that the percolation movement through a "neutral" landscape is strongly dependent upon the proportion occupied by favorable habitat. When a very large proportion of the landscape is occupied by favorable habitat the movement in any direction in the landscape is possible and we can indicate that percolation can happen in both directions in the landscape (typically N–S and E–W or, in our example, up–down and left–right). When the proportion of favorable habitat shrinks and becomes more fragmented the probability of percolation decreases. It is also important to note that with a "neutral" landscape approach the habitats are either favorable or not favorable. The quality of the favorable habitats is not assessed.

This critical threshold for the favorable class to allow percolation was further investigated under the percolation theory approach. The existence of these thresholds should be recognized by ecologists and land managers since it can be concluded from these analyses that, from the perspective of a species, small changes in the abundance of favorable habitats near the threshold level may cause significant effects on their populations. In fact, the free movement of organisms across landscapes is only theoretically possible if there is an infinite percolating cluster or network of favorable habitats[7].

6.2   Contagion Analysis and Percolation

Actual landscape patterns generally show significant departures from those randomly generated. Most processes produce "clumped" landscapes where cells of the same class are often contiguous to each other. This proximity between like cells is often a result of a contagious process. A way to measure contagion in actual landscape maps involves the calculation of the frequency of cell adjacencies.

For any class (j) we can determine q_{jj} as the observed proportion of cells of that class being adjacent to other cells of the same class (favorable or alike adjacencies). In the case of a random distribution of cells the probability that a cell of class j is adjacent to another cell of the same class is the probability that any cell is occupied by class j, that is, their proportion in the landscape p_j. The comparison between q_{jj} and p_j allows for some prediction to be made of the pattern of that class in the landscape.

We can now define a contagion index for class j (CONTAG$_j$), defined as

$$\text{CONTAG}_j = q_{jj}/p_j$$

when:

CONTAG$_j$ < 1 or $q_{jj} < p_j$, the class is dispersed (regularly distributed) through the landscape

CONTAG$_j$ = 1 or $q_{jj} = p_j$, the class is randomly distributed

CONTAG$_i$ > 1 or $q_{jj} > p_j$, the class is contagiously distributed (aggregated) in the landscape

Simple examples from Figure 6.6 with p_j = 6/16 = 0.375 illustrate the calculations. In the example on the left, the number of favorable adjacencies (left–right and up–down) is 1 + 1 = 2 and the number of unfavorable adjacencies is 3 + 3 = 6. The ratio between the favorable adjacencies (2) and the total number of adjacencies considered (2 + 6 = 8) results in a value of q_{ii} of 2/8 = 0.25. The value for the contagion index CONTAG$_j$ is 0.250/0.375 = 0.67, indicating that the class is regularly distributed in the landscape (CONTAG$_j$ < 1).

Similar calculations for the example in the right of Figure 6.6 indicate that there are 5 favorable adjacencies in a total of 8 adjacencies considered, resulting in a q_{ii} value of 5/8 = 0.625, and the value for the contagion index CONTAG$_j$ is 0.625/0.375 = 1.67, indicating that the class is contagiously distributed (aggregated) in the landscape (CONTAG$_j$ > 1).

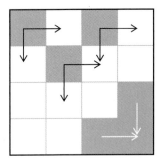 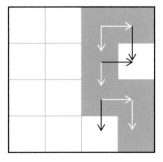

Figure 6.6 Two landscapes with 4×4 cells both having 6 cells in the favorable class j (in gray) and therefore the same proportion of that class in the landscape $p_j = 6/16 = 0.375$. Using only movements through edges (bond percolation) and only from left to right and from up to down we can count the favorable adjacencies (white arrows) and unfavorable adjacencies (blue arrows).

It is also important to note that, in spite of having the same proportion of favorable habitats, the second landscape shows the possibility of percolation (in one direction) whereas percolation is not possible in the example on the left. The spatial distribution of the class matters to potential movement.

It has been concluded in many studies that, besides the proportion of the class in the landscape (p_i), its distribution (measured by the degree of contagion) also affects the probability of percolation. Classes with low contagion (with numerous small patches) or with high contagion (with very large patches) tend to require proportions higher than 0.593 (the threshold level for random distributions) to achieve percolation[8,9].

A simple example can illustrate the influence of both the proportion of the favorable habitat and its distribution in the landscape (measured by the contagion index) in the possibility of percolation (Figure 6.7).

We can count adjacencies from the cells of a favorable habitat to adjacent cells at the right or at the bottom, excluding the edges of the landscape. From the first situation, in the upper left, with a proportion of favorable habitat of $p_j = 0.30$, it is possible to observe that, from the 55 edges analyzed, an organism moving from one favorable cell to its right or below would only find a cell with favorable habitat cells adjacent to it 4 times (1 top-to-bottom, 3 left-to-right). The ratio between the number of like adjacencies and the number of adjacencies analyzed in this case is, therefore, $q_{jj} = 4/55 = 0.07$. As this value is well below the proportion of favorable habitat ($p_j = 0.30$), the value of the contagion index ($\mathrm{CONTAG}_j = 0.24$) is much lower than unity, allowing for the conclusion that the favorable habitat is very highly dispersed through the landscape. A similar procedure was done for all the other 3 landscapes. A summary of the metrics analyzed is provided in Table 6.1.

From the same example it is possible to conclude that the possibility of percolation depends upon the combination of the proportion of favorable habitat (p_j) and its distribution. In the top two landscapes only a very strong contagion enables percolation at lower values of a favorable habitat, whereas at higher proportions of a favorable habitat a lesser degree of contagion is enough to allow percolation. In fact, if cells of a favorable habitat are highly aggregated, this can result in simple compact shapes that do not favor percolation.

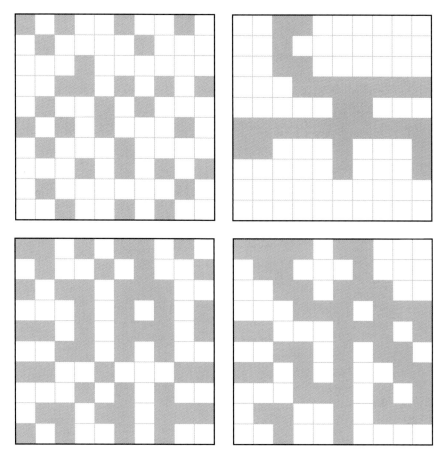

Figure 6.7 Landscapes represented by 10 × 10 grid cells with different proportions of favorable habitat (p_j = 0.30 in the two top landscapes and p_j = 0.50 in the two at the bottom) and with different types of spatial distributions (regularly dispersed in the two landscapes at the left and with some degree of contagion at the right).

Finally, it is important, as always, to consider scale. It could easily happen that a certain class can show a very high contagion at a certain scale and a regular (dispersed) distribution at another scale, as was demonstrated in detail with the distribution of points. Again, the important issue here is the use of the scale, or the grid, best suited to represent the objective of the analysis, as in the movement of organisms or disturbances.

There are different possibilities to look at the scale effect. So far, our analysis assumes that the organism, or disturbance, is restricted to moving between adjacent cells of suitable habitat (bond percolation). In this case, if the organism is to move through the landscape but remain within a certain class such as a forest, then the forest must occupy more than 59% (for a random landscape). However, if the organism can traverse a nonforested habitat (cells) and must find at least one occupied cell as it moves no more than k cells on the landscape, the probability R of finding at least one occupied cell is

$$R = 1 - (1 - p)^k$$

Table 6.1 Summary of metrics associated with changing landscape proportions of favorable habitat (p_j) and distribution pattern illustrated in the example of Figure 6.7, with n = number of patches, MPA = mean patch area, q_{jj} = proportion of like adjacencies (left to right and top to bottom), PLR = percolation from left to right, and PTB = percolation from top to bottom).

	Distribution of the class in the landscape	
Proportion of the favorable class in the landscape	Dispersed	Contagious
	$n = 26$	$n = 1$
	MPA = 1.2	MPA = 30
$p_j = 0.30$	$q_{jj} = 4/55 = 0.07$	$q_{jj} = 32/56 = 0.57$
	CONTAG$_j$ = 0.24	CONTAG$_j$ = 1.90
	PLR = No	PLR = Yes
	PTB = No	PTB = No
	$n = 18$	$n = 4$
	MPA = 2.8	MPA = 12.5
$p_j = 0.50$	$q_{jj} = 33/91 = 0.36$	$q_{jj} = 48/93 = 0.52$
	CONTAG$_j$ = 0.73	CONTAG$_j$ = 1.03
	PLR = No	PLR = Yes
	PTB = No	PTB = Yes

We can recall from percolation theory that if the organism could move across a random landscape it could find a favorable adjacent cell with a probability above a critical threshold $p_c = 0.593$. The same applies now for the probability of finding a favorable cell in the k neighborhood also being $R_c = 0.593$ for the random landscape. We can rearrange the equation to read

$$\ln(1 - R_c) = k \ln (1 - p_c)$$

Solving for k we have

$$k = \ln (1 - R_c)/\ln (1 - p_c) = -0.899/\ln (1 - p_c)$$

and solving for p_c we obtain

$$p_c = 1 - \exp (-0.899/k)$$

This equation, developed and used by various authors[10], allows the use of percolation theory of random landscapes with the appropriate scale, that is, the scale of the maximum cells of unsuitable habitat (k) that an organism is able to explore and continue to move across the landscape (Table 6.2).

It is easily seen from the above equations and table that organisms (or processes) that can traverse large distances (have large k values) can use a class that is sparsely distributed and still move through the landscape.

The approach shown is equivalent to changing the size of the grid. For example, in our examples, if the grid is now a group of 4 cells and a favorable habitat is considered as a group having at least one favorable cell, percolation would very easily occur in the landscapes where it was otherwise inhibited by excessive regularity.

Table 6.2 Minimum values for the proportion of the landscape occupied by a favorable habitat (the critical proportion p_c) to allow bond percolation for organisms that are able to move to k neighbor cells computed by the equation $p_c = 1 - \exp(-0.899/k)$.

k value	Critical proportion (p_c)
1	0.593
4	0.201
9	0.095
16	0.055

Ecological implications of this relationship are numerous. Firstly, the requirements for movement through landscapes are very dependent on scale and that has to be related to the organism or process of interest. Secondly, the proportion of favorable habitat plays an important role in the likelihood of percolation. Thirdly, the spatial distribution or configuration of the cells, as measured by their contagion index, can be very important. Finally, it should again be emphasized that the notion of favorable habitat is absolutely dependent upon the organism of interest. Habitats that can be used as a conduit for some organisms are often a barrier for the movement of others. The notions of habitat, conduit, barrier, and filter discussed in the chapter of linear elements are very useful to recall at this point.

6.3 Resistance Surfaces

Applications of percolation theory require a landscape that is classified into a habitat that is suitable or unsuitable for movement of an organism. In reality, this is a simplified view of the relationship between the organism and the landscape it is attempting to move through. The likelihood of an organism moving through different habitat patches can be compromised by the "cost" of moving through that landscape, where the cost can be represented by energy expenditure or exposure to threats. One approach recently developed to quantify the gradual ability of an organism to move across habitat patches of varying quality is the idea of representing the landscape by a "resistance surface".

Development of resistance surfaces can be accomplished by considering a number of environmental variables and applying numeric values to the environmental gradients where a low value indicates ease of movement and a high value indicates resistance or cost to the movement[11]. A low resistance value may represent a preferred habitat or suitable movement corridors while very high resistance or cost values may be applied to a road, a river, or a human-dominated area that is extremely difficult or dangerous for the organism to traverse.

The development of a resistance surface has been described in four steps[12]. In the first step the modeler selects environmental variables thought to impact the movement of the organism of interest. The environmental variables are represented by geospatial data layers developed by the modeler or available via other sources. The layers should be adjusted to the appropriate scale for the analysis. In the second step, biological data for the organism of interest are obtained. The biological data can consist of field observations, GPS locations, or can be based on expert opinion. In step three the environmental and biological data are combined to create possible resistance surfaces and in step four the proposed resistance surfaces are compared and one may be selected as the final resistance surface.

A variety of environmental variables have been noted in the literature in the creation of resistance surfaces, including topographic variables, land cover, land use, human population density, climate variables, and reflectance values from satellite imagery or aerial photography. Linear features such as roads, streams, and riparian corridors can also be incorporated in a resistance surface, but particular care must be taken when including such narrow features in a raster-based resistance surface. Consider a resistance surface composed of two habitats (Figure 6.8): traversable (yellow) and habitat that is nontraversable or difficult to traverse (blue). If site perkolation (the eight-cell rule) is used to determine connectivity, in theory the organism could move through the nontraversable habitat at the pixel corners. This phenomenon has been called "cracks" in the resistance surface[13]. Such cracks can be addressed by buffering linear features to make them slightly thicker and thereby filling in the cracks via the spatial methodologies described[14,15].

The developed resistance surface can now be used to model the least-cost paths of an animal's movement from one point to another across the landscape. An example identifying movement corridors for lynx (*Lynx pardinus*) on the Iberian Peninsula using resistance surfaces and least-cost modeling[16] is shown in Figure 6.9. GPS locations from animals were fitted to resource selection functions with explanatory variables including topography, twelve land cover types, and road variables (high and low traffic). The resistance surface was then assigned a minimum value of 0 for low resistance to movement and 100 for maximum resistance to movement (Figure 6.10). Corridors requiring the lowest travel cost for the lynx were identified using least-cost path modeling in GIS.

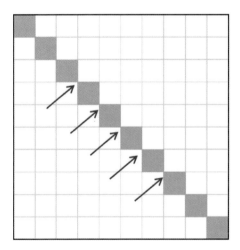 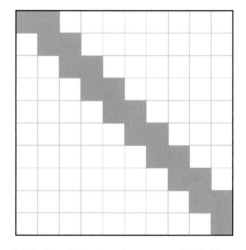

Figure 6.8 Resistance surfaces composed of traversable habitat (yellow) and nontraversable habitat (blue). The figure to the left exemplifies cracks in the resistance surface where the organism can move through a nontraversable area. The figure to the right represents a solution where the cracks in the resistance surface are eliminated and the linear feature is truly nontraversable.

Figure 6.9 The Iberian lynx (*Lynx pardinus*) is an endangered species native to southern Europe's Iberian Peninsula. The Iberian lynx habitat has been lost and fragmented. Projects to understand habitat relationships of lynx have been launched. *Source:* Painting by Luisa Nunes for the logo of the POSEUR project MODELYNX, http://www.isa.utl.pt/ceabn/uploads/img/gestao_vida_silvestre/poseur/ Logo_Lince.jpg.

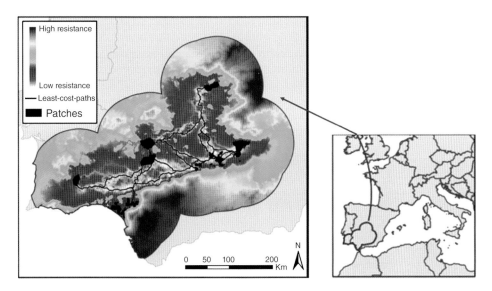

Figure 6.10 Stable populations and reintroduction areas (in black) identified for the Iberian lynx by the research project IBERLINCE, overlayed with a resistance surface ranging from low resistance (0; white/ pink colors) to high resistance (orange/brown colors). Proposed linkages between populations are represented by black lines. *Source:* Blazquez-Cabrera, S., Gastón, A., Beier, P., *et al.* (2016) Influence of separating home range and dispersal movements on characterizing corridors and effective distances. *Landscape Ecology*, 31, 2355–2366.

Three different ways of creating the resistance surface based on only home range estimation were tested, including dispersed animal locations in the model and a combination of the two. The scientists concluded that resistance surfaces developed from only the home range data resulted in lower connectivity estimates and emphasized the importance of including dispersed individuals in the development of the resistance surface.

Spatial analysis tools have been developed to model movement across landscapes using resistance surfaces. For example, the Guidos Toolbox[17] provides GIS tools to create cost surfaces, least-cost mapping, and assess connectivity between patches. The ArcGIS[18] software also provides tools for creating raster resistance surfaces and distance tools for creating cost surfaces and estimating the least-cost path between locations.

Thus resistance surfaces are a method to more realistically estimate the movement across landscapes of variable composition and habitat characteristics than the methods that classify habitat into either a suitable or an unsuitable habitat. However, resistance surfaces require more information and assumptions about the interactions between the organism and the landscape. Resistance surfaces may also have applications in describing the movement of other landscape components such as nutrients and disturbances.

6.4 Example of Percolation Movements Through Landscapes

In this section we provide an example to illustrate the use of percolation and contagion analysis on movements of animals through landscapes. In this example we use two small rodents requiring different types of habitats in western North America (Figures 6.11 and 6.12). The pinyon deermouse (*Peromyscus truei)*, a small rodent of arid and semi-arid zones of western North America, is found most often in pine and juniper woodlands[19]. It can be considered to be a woodland obligate. On the other hand, the pygmy rabbit (*Brachylagus idahoensis*) is a small rabbit that uses mainly sagebrush areas as habitat, feeding on grasses and leaves in summer and on bark and sagebrush twigs in winter. It can be considered a sagebrush obligate.

Landscapes where these two species occur are the subject of great changes. The images shown in Figure 6.13 represent landscapes from the Owhyee Mountains in western USA

Figure 6.11 Mature western juniper (*Juniperus occidentalis*) woodland (left), habitat of the pinyon deermouse (*Peromyscus truei*) (right). *Sources:* Stephen Bunting (left) and http://www.nps.gov/band/naturescience/id-of-mammals.htm (right).

Figure 6.12 The big sagebrush (*Artemisia tridentata*) steppe (left), habitat of the pygmy rabbit (*Brachylagus idahoensis*) (right). *Sources:* Stephen Bunting (left) and Courtesy of Janet Rachlow, University of Idaho (right).

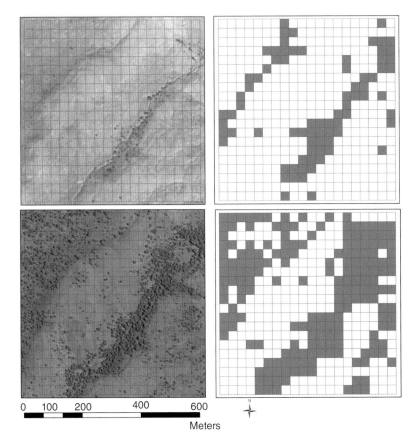

0   100   200   400   600
Meters

Figure 6.13 Images of the Owhyee Mountains in 1940 (top left) and in 2004 (bottom left) transformed to the right in 20 × 20 binary landscapes in a grid system (30 m × 30 m cells), showing sagebrush cells (white) and juniper cells (gray). *Sources:* Top left image: https://earthexplorer.usgs.gov/; bottom left image: Digital Orthorectified Aerial Imagery Series of Idaho (2004, 1 meter, natural color), https://insideidaho.org.

Table 6.3 Summary of landscape metrics associated with changing landscape proportions of habitat classes and distribution pattern illustrated in the example shown in Figure 6.13, where p_j = proportion of habitat j, q_{jj} = proportion of like adjacencies (left to right and top to bottom), CONTAG$_j$ = contagion index, PLR = percolation from left to right, and PTB = percolation from top to bottom).

	Class in the landscape	
Image	Class 1 Juniper (gray cells)	Class 2 Sagebrush (white cells)
1940	$p_1 = 78/400 = 0.20$ $q_{11} = 79/148 = 0.53$ CONTAG$_1$ = 2.74 PLR = No PTB = No	$p_2 = 322/400 = 0.81$ $q_{22} = 541/612 = 0.88$ CONTAG$_2$ = 1.10 PLR = Yes PTB = Yes
2004	$p_1 = 189/400 = 0.47$ $q_{11} = 248/361 = 0.69$ CONTAG$_1$ = 1.45 PLR = No PTB = Yes	$p_2 = 211/400 = 0.53$ $q_{22} = 292/399 = 0.73$ CONTAG$_2$ = 1.39 PLR = No PTB = Yes

show the expansion of juniper woodlands in areas previously occupied by different types of sagebrush steppe communities. From the perspective of the two species of interest the landscape can be classified as favorable and unfavorable habitats. As both species have small home ranges, the grain used in the analyses was based on 30 m × 30 m grids.

For the pygmy rabbit the favorable habitat is the sagebrush (white squares) whereas for deermouse the favorable habitat is the juniper woodland (gray squares). We have therefore to calculate metrics that are meaningful for each species (Table 6.3).

From this example it is possible to conclude that the changes in the proportion of classes in the landscape and their spatial arrangement are likely to have consequences for the populations of the two species of rodents and in their movements through the landscapes.

For the pinyon deermouse, the increase in juniper habitat was beneficial, already allowing some percolation at a proportion of suitable habitat of 47%, due to the contagious pattern that created corridors. For the pygmy rabbit, the changes were not favorable, because of loss of sagebrush steppe habitat and restricted movement, as shown by the lack of percolation from left to right.

This example is particularly illustrative as it highlights the fact that analysis of the landscape and the movements through it have to be made in relation to the specific organism or process of interest.

Key Points

- Movement of organisms or processes through landscapes is an important aspect of understanding the ecology of those landscapes.
- Percolation theory has been applied to quantify movement of organisms through landscapes composed of favorable and unfavorable patches. When the number of favorable

habitat patches increases, it becomes increasingly likely that the organism can move through the landscape.

- It is important to distinguish between bond and site percolation. In bond percolation the movement can only occur to four neighboring cells through the bonds of adjacent cells (edges), while in site percolation the movement can also occur through sites or corners (the vertices of the cells) with each cell having eight neighbors.
- Percolation through a landscape becomes increasingly easy when the proportion of favorable patches reaches a proportion of 0.593 for bond percolation and 0.50 for site percolation.
- Contagion is a measure of how "clumped" or aggregated landscape units of the same class are.
- The ability of an organism to move through a landscape depends both on the proportion of favorable habitat and the aggregation of the habitat patches.
- As landscapes change through successional or other processes, movement through the landscape can become more favorable for one species while less favorable for others.

Endnotes

1 Stauffer, D. (1985) *Introduction to Percolation Theory*, Taylor and Francis, London.

2 Milne, B.T. (1992) Spatial aggregation and neutral models in fractal landscapes. *American Naturalist*, 139, 32–57.

3 Gardner, R.H., Milne, B.T., Turner, M.G., and O'Neill, R.V. (1987) Neutral models for the analysis of broad-scale landscape pattern. *Landscape Ecology*, 1, 19–28.

4 Pearson, S.M. and Gardner, R.H. (1997) Neutral models: Useful tools for understanding landscape patterns, in *Wildlife and Landscape Ecology: Effects of Pattern and Scale* (ed. J.A. Bissonnette), Springer Verlag, New York, NY.

5 Stauffer, D. (1985) *Introduction to Percolation Theory*, Taylor and Francis, London.

6 Gardner, R.H. and O'Neill, R.V. (1991) Pattern, process, and predictability: The use of neutral models for landscape analysis, in *Quantitative Methods in Landscape Ecology* (eds M.G. Turner and R.H. Gardner), Springer-Verlag, New York, NY.

7 Milne, B.T. (1991) Lessons from applying fractal models to landscape patterns, in *Quantitative Methods in Landscape Ecology – Then Analysis and Interpretation of Landscape Heterogeneity* (eds M.G. Turner and R.T. Gardner), Springer-Verlag, New York, NY.

8 Gardner, R.H. and O'Neill, R.V. (1991) Pattern, process, and predictability: the use of neutral models for landscape analysis, in *Quantitative Methods in Landscape Ecology* (eds M.G. Turner and R.H. Gardner), Springer-Verlag, New York, NY.

9 Hargis, C.D., Bissonette, J.A., and David, J.L. (1998) The behavior of landscape metrics commonly used in the study of habitat fragmentation. *Landscape Ecology*, 13, 167–186.

10 O'Neill, R.V., Krummel, J.R., Gardner, R.H., *et al.* (1988) Indices of landscape pattern. *Landscape Ecology*, 3, 153–162.

11 Zeller, K.A., McGarigal, K., and Whiteley, A.R. (2012) Estimating landscape resistance to movement: A review. *Landscape Ecology*, 27, 777–797.

12 Zeller, K.A., McGarigal, K., and Whiteley, A.R. (2012) Estimating landscape resistance to movement: A review. *Landscape Ecology*, 27, 777–797.

13 Adriaensen, F., Chardon, J.P., De Blust, G., *et al.* (2003) The application of "least-cost" modeling as a functional landscape model. *Landscape and Urban Planning*, 64, 233–247.

14 Rothley, K. (2005) Finding and filling the "cracks" in resistance surfaces for least-cost modeling. *Ecology and Society*, 10, 4; [online] URL: http://www.ecologyandsociety.org/vol10/iss1/art4/.

15 Theobald, D. (2005) A note on creating robust resistance surfaces for computing functional landscape connectivity. *Ecology and Society*, 10, r1; [online] URL: http://www.ecologyandsociety.org/vol10/iss2/resp1/.

16 Blazquez-Cabrera, S., Gastón, A., Beier, P., *et al.* (2016) Influence of separating home range and dispersal movements on characterizing corridors and effective distances. *Landscape Ecology*, 31, 2355–2366.

17 Vogt, P. and Riitters, K. (2017) GuidosToolbox: Universal digital image object analysis. *European Journal of Remote Sensing*, 50, 352–361; [online] URL: http://forest.jrc.ec.europa.eu/download/software/guidos.

18 ESRI (2017) ArcGIS Desktop: Release 10.4 Redlands, CA: Environmental Systems Research Institute; [online] URL: http:www.esri.com.

19 Hammond, D.B. and Yensen, E. (1982) Differential microhabitat utilization in *Peromyscus truei* and *Peromyscus maniculatus* in the Owyhee Mountains, Idaho. *Journal Idaho Academy Science*, 18, 49–56.

7

Landscape Composition, Diversity, and Habitat Selection

Diversity is a key concept for all ecological studies. At the landscape scale diversity often refers to the composition of the landscape, that is, the number and the proportions of the various classes or types of habitat present in that landscape. However, diversity is also often a measure of the variety of species and their proportions. Biodiversity is now defined[1] as "the variety of life on Earth, including all organisms, species, and populations, the genetic variation among these, and their complex assemblages of communities and ecosystems. It also refers to the interrelatedness of genes, species, and ecosystems and in turn, their interactions with the environment".

Because of the complexity of the issue and the possible difficulties and confusion in the use of the term Diversity, it is important to start by presenting the various measurements of diversity and then the relationships between these various types of diversity, from the species diversity of habitats and landscapes to the habitat selection and habitat use diversity of the species.

7.1 Measurements of Diversity

Early civilization humans have attempted to describe the diversity of life (Figure 7.1). Biodiversity can be considered at different levels, from genes to species and from communities and habitats to landscapes. Interestingly, the same diversity measurements apply along the different levels of the ecological hierarchy. However, for simplicity, the initial measurements of diversity focused on the number of species found within a given area.

Species richness (S), the first described index of diversity, is the count of the total number of species present within the area of study. In actuality, as numeration of all species is a virtual impossibility, species richness normally includes only a small portion of the total biota, such as vascular plants or breeding birds.

Species richness is a simple and easily comprehensible expression of diversity, and we already know from island biogeography and metapopulation theories that species richness depends on the group of species studied, the extent and nature of the study area, and its proximity to other similar areas.

Applied Landscape Ecology, First Edition. Francisco Castro Rego, Stephen C. Bunting, Eva Kristina Strand and Paulo Godinho-Ferreira.
© 2019 John Wiley & Sons Ltd. Published 2019 by John Wiley & Sons Ltd.

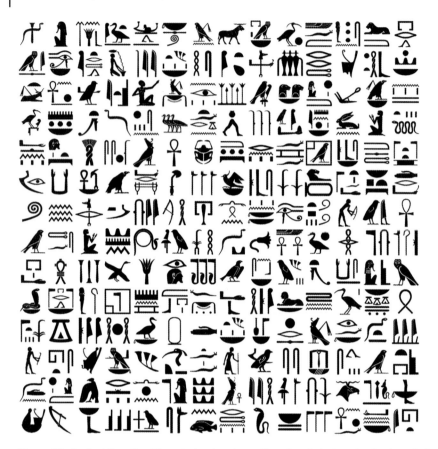

Figure 7.1 Ancient Egyptian hieroglyphs showing a diversity of plant and animal symbols. *Source:* http://previews.123rf.com/images/ddart/ddart1105/ddart110500001/9532981-Ancient-Egyptian-hieroglyphs-Stock-Vector-egyptian-egypt-hieroglyphics.jpg.

However, the use of the number of species implies that all species are treated equally regardless of their abundance. Species with a very low number of individuals in the area are considered similarly to those with a large number of individuals, as the relative abundances of species are not taken into account. However, it is known that the abundance of species is of great ecological importance, providing, for example, a more sensitive measure of environmental diversity than species richness alone[2].

If abundance is taken into account in measuring the species diversity, an area with all the species of equal abundance should be considered more diverse than another area with the same number of species but with some species common and others rare. Beginning in the 1940s there was increased interest to develop an index of complexity that included both an area's species richness and the abundance of those species.

One of the first statistical analyses taking into account species abundance was proposed in 1943 by Fisher and others[3] using a log-series distribution to describe species abundance distributions in large moth collections (Figure 7.2).

Figure 7.2 The English statistician and biologist Sir Ronald Fisher (1890–1962) studied the statistical distributions of various collections of moths. *Source:* FortyTwo Evolution, http://www.42evolution.org/ronald-a-fisher/(left) and https://commons.wikimedia.org/wiki/File:Assorted_Moths_(Lepidoptera)_in_the_University_of_Texas_Insect_Collection_(22281153644)_(cropped).jpg (right).

Fisher found that several moth collections studied followed a log-series distribution. Under this distribution the expected number of species with x individuals $E(x)$ is

$$E(x) = \alpha \beta^x / x$$

where the parameters α and β could be estimated.

The parameter α is therefore independent of the sample size and has the attractive property that it may be used as a diversity statistic[4]. However, this approach has the limitation that the data should follow a log-series distribution, which is often not the case.

Two essentially different ways of incorporating abundance into diversity indices have been developed. First, the distribution of relative abundance of species can be fitted by statistical distributions that may have parameters which can be used as diversity indices as the parameter α of the logarithmic series above.

A second approach is independent of any hypothetical distribution of relative abundance and takes a number of different alternatives, based on the use of the values of p_j, the proportion of the jth species in a collection of species. Nearly all diversity indices are based on estimates of the relative abundance of the species:

$$p_j = N_j / N$$

where N_j is the number of individuals of the jth species in the total of N individuals of the S species present in the area.

In this second approach, instead of fitting distributions, diversity began to be described by indices. Important contributions to diversity indices came in the late 1940s from works such as that on communication and information theory by Claude Shannon[5,6] and those of Edward Simpson[7] (Figure 7.3).

Figure 7.3 Claude Shannon (1916–2001), the American mathematician known as "the father of information theory" (left). The British statistician Edward Simpson (born in 1922) (right). *Sources:* http://www.bloomberg.com/bw/articles/2014-12-04/information-theory-intellectual-foundation-for-the-digital-age (left) and http://onlinelibrary.wiley.com/doi/10.1111/j.1740-9713.2010.00424.x/full (right).

The indices proposed by these two scientists, together with species richness and the simple index later proposed by Berger and Parker[8], constitute the most commonly used measures of species diversity of an area.

From the mathematical point of view we can define abstractly a stochastic process that generates a sequence of symbols with given probabilities, as did Shannon[9]. Suppose that we consider a long message (community in ecology) of N symbols (individuals) with five letters A, B, C, D, and E (species), which are chosen (or found) with probabilities p_A, p_B, p_C, p_D, and p_E, with successive choices being independent. The sequence of symbols will contain, with high probability, about $p_A N$ occurrences of the first symbol, $p_B N$ occurrences of the second, etc. Hence the probability of this particular message will be

$$p = p_A{}^{p_A N} p_B{}^{p_B N} p_C{}^{p_C N} p_D{}^{p_D N} p_E{}^{p_E N}$$

The inverse of this probability $(1/p)$ is therefore a measure of the diversity of possible sequences. As the diversity of possible sequences is a function of the length of the sequence N it is convenient to take the log transformation:

$$\ln (1/p) = \sum \left[(p_j N) \ln (1/p_j) \right] = -N \sum \left[(p_j) \ln (p_j) \right]$$

It is then possible to have a measure of diversity that is independent of the length of the sequence:

$$SH = \ln (1/p)/N = -\sum \left[(p_j) \ln (1/p_j) \right]$$

This index (Shannon's SH) was the initial diversity measure and still one of the more frequently used indices in ecology.

This index has also been termed the Shannon–Weaver index, as Warren Weaver (1894–1978), another American scientist, co-authored with Shannon in 1949 a landmark book[10] that made the information theories accessible to nonspecialists, *The Mathematical*

Theory of Communication. The same index can also be named the Shannon–Wiener index, recognizing the contribution of another American mathematician, Norbert Wiener (1894–1964), who also developed important advancements on information theory independently of Shannon during World War II.

At the same time, Simpson[11] developed his index, which used the same type of data, species richness, and abundance. The Simpson index (SI) measures the probability that two individuals randomly selected from a sample will belong to the same species:

$$SI = \sum p_j^2$$

It is intuitive that the more diverse the sequence the less probable it is that two individuals randomly selected will belong to the same species. Therefore, transformations of the original index as $1 - SI$ or $1/SI$ are commonly used and are also referred to as the Simpson diversity index.

Finally, the Berger–Parker index (BP) equals the maximum p_j value, that is the proportion of the most abundant species[12], which is also inversely related to diversity:

$$BP = \max(p_j)$$

During the decades following the initial work of Shannon and Simpson more than 50 indices were developed using many variations on the combination of species richness and abundance. They have been extensively reviewed and compared by many authors[13,14].

However, it was Hill[15] who, in 1973, concluded that all these indices could be transformed into a set of "diversity numbers", which will be referred to here[16] with the notation H_a (Hill).

The general equation for the diversity numbers of the Hill series (H_a) is

$$H_a = \left(\sum p_j^a\right)^{1/(1-a)}$$

For $a = 0$, H_0 is equal to S, the total number of species in the sample (species richness). For $a = 1$, H_1 is equivalent to exp (SH), where H' is the Shannon index; $H_1 = \exp\left\{-\sum_{j=1}^{s}(p_j))\ln(p_j)\right\}$.

For $a = 2$, H_2 is equivalent to 1/SI, where SI is the Simpson index; $H_2 = 1/\sum_{j=1}^{s} p_j^2$.

For a = +∞, $H_{+\infty}$ is equivalent to 1/BP, where BP is the Berger–Parker index; $H_{+\infty} = 1/\max(p_j)$.

We will also refer to H_1 as the modified Shannon index as it is the exponential of the original Shannon Index: $H_1 = \exp(SH)$. Similarly, we will also refer to H_2 as the modified Simpson index as it is the inverse of the original Simpson index: $H_2 = 1/SI$. We can therefore evaluate this Hill series index for any value using the general equation, reflecting a gradient of diversity indices that change emphasis from the rarest species at one end to the most common species at the other end.

In addition to showing a sequence of emphasis on rare species (S) to place emphasis on the most common species ($H_{+\infty}$), the diversity numbers of Hill also have the advantage of being expressed in much more "natural units" as they indicate an equivalent number of species, that is, the number of species containing the same number of individuals that would yield the same diversity value.

Obviously, if all species have the same abundance all the diversity numbers H_a would be equal to the total number of species present (S), which is the upper limit for all diversity numbers. In special cases the different diversity numbers can be equal but the general sequence of the Hill series is the following:

$$H_0 > H_1 > H_2 > H_{+} \infty$$

This indicates that H_1, the modified Shannon index, is closer to the species richness H_0 and consequently more sensitive to changes in rare species than H_2, the modified Simpson index, which is closer to $H_{+\infty}$ and therefore more sensitive to changes in abundance of the more common species. As the value of Hill's power constant a increases, the diversity number comes to depend more and more on the common species and less and less on the rare. In general, the more even the proportions, the less variable H_a will be over the range of a.

Several studies[17] indicate that H_1 is the most sensitive index to environmental changes and its strength from support of information theory suggests it is the single best option in diversity studies. However, since different indices indicate different aspects of species abundance, it is often useful to use several indices simultaneously.

As Hill points out, the concept of evenness can now be thrown into a clearer light. The ratio between different diversity numbers is related to evenness. As the maximum diversity value is species richness ($H_0 = S$) all of the other diversity indices can be related to this maximum value and we can have three different evenness measures ($E_a = H_a/H_0 = H_a/S$):

$$E_1 = H_1/S, \quad E_2 = H_2/S, \quad \text{and} \quad E_3 = H_3/S$$

In general terms:

$$\text{Diversity } H_a = \text{Richness } S \times \text{Evenness } E_a$$

These approaches to the measurement of diversity based on proportions of the different species are often based on the number of individuals. However, species abundance can also be expressed in terms of biomass, density, or coverage (basal or canopy). It is evident that the choice of parameter used will affect the relationship between species and the index value itself. Therefore, species diversity expressed in terms of biomass will be different than if it was expressed in terms of density.

7.2   Species Diversity of Habitats and Landscapes

As in the other aspects of landscape ecology, scale is also very important when measuring diversity (as seen in the discussion of patch size in Chapter 4), and species diversity has been applied from small areas (individual habitats) to large areas (landscapes, regions, or the entire world) (Figure 7.4).

Species diversity can therefore be evaluated in points (as in common bird surveys) to regions or the entire world. Whittaker[18,19], in his important work in the 1970s also distinguishes various levels, referring to the species diversity of points (*point* diversity),

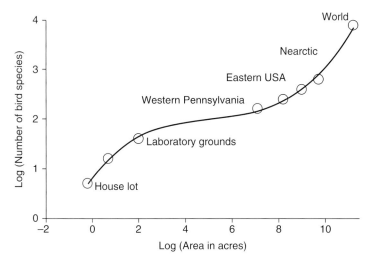

Figure 7.4 Number of bird species as a function of area (log scales) from a very small area to the entire world showing the importance of scale. Adapted from various sources: (1) Preston, F.W. (1960) Time and space and the variation of species. *Ecology*, 41, 785–790; (2) Rosenzweig, M.L. (1995) *Species Diversity in Space and Time*, Cambridge University Press, Cambridge, MA; (3) Ladle, R.J. and Whittaker, R.J. (eds) (2011) *Conservation Biogeography*, Blackwell, Oxford, UK.

to that of communities (or habitats) as *alpha* diversity, that of several communities (landscapes as a mosaic of habitats) as *gamma* diversity, and that of a region as *epsilon* diversity.

In this book we are basically interested in two types of species diversity that MacArthur[20] refers to as within and between habitats. Here we define two levels of species diversity:

a) The Species Diversity of a Habitat (SDH), referring to a homogeneous area (habitat), corresponding to the definition of *alpha* diversity in Whittaker's terminology.
b) The Species Diversity of a Landscape (SDL), referring to heterogeneous areas (landscapes) as a mosaic of different habitats, corresponding to *gamma* diversity in the terminology of Whittaker.

According to the ecological hierarchy already discussed, the plant and/or animal communities that are the basis of the species diversity of a specific habitat (SDH) are also included at a higher level in the species diversity of the landscape (SDL) where this habitat is included.

Because of this hierarchy the species diversity of a landscape (SDL), if measured as its richness (the number of species present in the landscape) is always equal or greater than the richness of any of its habitats. All species included in one habitat are also included in the landscape where that habitat occurs.

However, when using indices that include abundance and not only presence–absence data this situation may change. We will discuss these issues in the following example.

A good example of the analysis of species diversity in habitats (SDH) and species diversity in landscapes (SDL) can be illustrated by the calculation of indices of diversity of

Figure 7.5 The four pigeons that inhabit the island of São Tomé. From left to right and top to bottom: São Tomé maroon pigeon (*Columba thomensis*), lemon dove (*Columba simplex*), green pigeon (*Treron sanctithomae*), and bronze-naped pigeon (*Columba malherbii*). *Sources:* Courtesy of César Garcia, Museu Nacional de História Natural, Lisboa, Portugal, http://cba.fc.ul.pt/members/cesar_garcia.php (upper left); by Alandmanson (own work) [CC BY-SA 4.0 (https://creativecommons.org/licenses/by-sa/4.0)], via Wikimedia Commons (upper right); by Albert Eckhout – Dante Martins Teixeira: Os quadros de aves tropicais do Castelo de Hoflössnitz na Saxônia e Albert Eckhout (ca. 1610–1666), artista do Brasil Holandês. Rev. Inst. Estud. Bras., 2009, Nr. 49, S. 67–90. ISSN 0020-3874. (Online Version), Public Domain, https://commons.wikimedia.org/w/index.php?curid=27087788 (lower left); by John Gerrard Keulemans – Bulletin of the Liverpool Museums, Public Domain, https://commons.wikimedia.org/w/index.php?curid=11444405 (lower right). On the right a landscape image of the island. *Source:* https://en.wikipedia.org/wiki/Geography_ of_São_Tomé_and_Príncipe.

pigeon species in the island of São Tomé, in the Gulf of Guinea (Figures 7.5 and 7.6). This island is an important center of bird diversity and endemism, and three of the four pigeons that inhabit the island are single island endemics maroon pigeon (*Columba thomensis), lemon dove (C. simplex)*, and bronze-naped pigeon (*Treron sanctithomae*) and the other (*Columba malherbii*) is only found in two other islands and Equatorial Guinea. Studies on the conservation of these forest pigeons make it possible to evaluate the diversity of pigeons in the different habitats of the island.

In Table 7.1 we calculate the proportion (p_{jk}) of species j in habitat k by dividing the number of counts in each cell (N_{jk}) by the total number of counts in each column ($N_{\cdot k}$), that is, the total number of individuals of all species counted in that habitat. For example, the proportion of species 1 (*Columba thomensis*) in the total bird counts in habitat 1, old-growth forests, can be indicated as p_{11} and computed as $p_{11} = N_{11}/N_{\cdot 1} = 15/205 = 0.073$.

We can now use these values of the proportions p_{jk} to evaluate the species diversity of habitats and the species diversity of the total island (as a landscape) using the equations of Hill for the equivalent number of species. The results are shown in Table 7.2.

For the most simple case, the nonforested habitat, we have only found two species ($H_0 = 2$). The other diversity indices can be computed for this habitat as

$$H_1 = \exp\left\{ - \left[8/102 \ln 8/102 + 94/102 \ln 94/102 \right] \right\} = 1.32$$

$$H_2 = 1/\left[(8/102)^2 + (94/102)^2 \right] = 1.17$$

$$H_{+\infty} = 1/(94/102) = 1.09$$

Figure 7.6 The general distribution of the four main types of habitat of the island of São Tomé (from left to right and top to bottom: old growth, secondary and shade forests, and nonforested habitats). *Source:* Adapted from Carvalho, M. (2014) *Hunting and Conservation of Forest Pigeons in São Tomé (West Africa)*, PhD Dissertation, Instituto Superior de Agronomia, Universidade de Lisboa.

Table 7.1 Counts of the different bird species on transects made in the wet season in the various types of habitat on the island of São Tomé. In parentheses we show the proportion of birds of each species in the total number of birds counted.

		Type of habitat				
Bird species	Common names English (and Portuguese)	Old growth forest	Secondary forest	Shade forest	Non-forested	Total count per species (N_j)
Columba thomensis	Maroon pigeon (Pombo-do-mato)	15 (0.073)	1 (0.002)	0 (0.000)	0 (0.000)	16 (0.013)
Columba simplex	Lemon dove (Muncanha)	90 (0.439)	220 (0.346)	71 (0.284)	8 (0.078)	389 (0.326)
Treron sanctithomae	São Tomé green pigeon (Céssia)	69 (0.337)	59 (0.093)	16 (0.064)	0 (0.000)	144 (0.121)
Columba malherbii	Bronze-naped pigeon (Rola)	31 (0.151)	355 (0.559)	163 (0.652)	94 (0.922)	643 (0.539)
Total count per habitat ($N_{.k}$)		205	635	250	102	$N = 1192$

Adapted from Carvalho, M., Fa, J.E., Rego, F.C., *et al.* (2015) Factors influencing the distribution and abundance of endemic pigeons in São Tomé Island (Gulf of Guinea). *Bird Conservation International, 25*, 71–86.

Table 7.2 Results of the evaluation of species diversity measured as the equivalent number of species obtained with the different indices of the Hill series for the different types of habitat and for the whole island.

	Types of habitat				
Species diversity (equivalent number of species)	Old growth forest	Secondary forest	Shade forest	Non-forested	Total island
H_0 Species richness	4	4	3	2	4
H_1 Modified Shannon index	3.34	2.52	2.25	1.32	2.75
H_2 Modified Simpson index	2.99	2.27	1.96	1.17	2.43
$H_{+\infty}$ Inverse of Berger–Parker	2.28	1.79	1.53	1.09	1.85
Average H	3.15	2.64	2.19	1.39	2.76

These values can also be displayed graphically showing the decreasing values of the Hill series for each habitat and allowing for the comparison between habitats and the total island (Figure 7.7).

It can easily be seen from the graph of Figure 7.7 that all indices indicate the higher diversity of pigeons in the habitat of old-growth forests, where the four species were observed (high richness) and where all species were found in relatively high proportions (high evenness). The steeper the slope of the line the more uneven is the distribution.

For secondary forests we also observed the four species, but one of them with just one count. This explains the abrupt decrease of the diversity numbers from 4 in H_0 to values between 2.5 and 2.0 for H_1 and H_2. In fact, as two of the species are much less represented

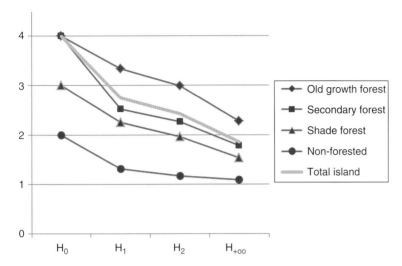

Figure 7.7 Species diversity evaluated as the number of equivalent species of pigeons (*Y* axis) for the Hill series (*X* axis) for the different types of habitat and for the total island of São Tomé (different markers and lines), illustrating the results presented in Table 7.2.

than the other two, the equivalent number of species H_1 and H_2 is not much higher than two.

All indices show that the shade forest has lower diversity than old-growth and secondary forests but is higher than nonforested habitats, where only two species were found but with one much less represented than the other, showing that the values for the equivalent number of species for the other indices are only slightly above unity. In fact, only one species is well represented.

When calculating the same indices of the Hill family for the whole island (all habitats together) the values fall below those of the habitat with the highest diversity (old-growth forests) and that with a higher number of individuals (the secondary forest).

The situation of having the diversity for the whole landscape lower than that of one of its habitats does not occur when using species richness. In that case the number of species in the landscape is at least the same as that of one of its habitats. However, when using indices that are not only based on presence–absence data but on quantities (as counts), this situation may occur. In fact, this case is a good illustration. All species are present and well represented in old-growth forests and all other habitats are significantly less diverse and do not contribute with new species to the overall species pool. In this case the species diversity of the whole landscape (SDL) is equal to or less than the species diversity of the most diverse habitat (SDH).

For conservation of the diversity of pigeons in the island, a landscape with only old-growth forests would be the optimal solution.

7.3 The Habitat Use Diversity of a Species

In the above section we discussed the diversity of species in habitats and landscapes. The measures of diversity were in number of species (richness) or the equivalent number of species from the indices of the Hill series.

In this section we use a parallel approach to measure the diversity of habitats used by a single species (habitat use diversity of a species (HUDS)) and the diversity of the use of the habitats by all the species considered (global habitat use diversity (GHUD)). In these cases the measurements of diversity are made in a number of habitats.

We will start with the habitat use diversity of a species (HUDS). The diversity of habitats that a species occupies is a function of the habitat breadth of the species, the diversity of habitats available in the landscape, and the habitat selection of the species.

It is known that different species typically use different habitats in different proportions. If a species uses only one habitat it is because it is the only habitat available or because it is an extremely specialized species with a very narrow ecological niche and therefore completely dependent upon the conservation of that habitat. These species are generally at risk of local extinction if that habitat is threatened. If the species is found in several habitats (a large HUDS) it is because there are several habitats available in the landscape and because the species is a generalist.

In the simpler approach the habitat use diversity of a species can be simply measured by the number of habitats where that species occurs. However, habitat use diversity of a species is generally best measured by an index that incorporates quantities and not only presence–absence. There is therefore the opportunity to use the equivalent number of habitats as measured by any of the indices of the Hill series.

For instance, the habitat use diversity of species j (HUDS$_j$) can be measured using the modified Shannon index as

$$\text{HUDS}_j = \exp\left\{ -\sum [u_{kj} \ln (u_{kj})] \right\}$$

where u_{kj} is the proportion of individuals of species j that were observed using habitat k.

We can use the same example of the birds in the island in São Tomé[21] to exemplify the calculations. We can now calculate the proportion (u_{kj}) of habitat k on the distribution of species j by dividing the number of counts in each cell (N_{jk}) by the total number of counts in each row, that is, the total number of individuals of species j observed (N_j). For example, the proportion of habitat 1 (old-growth forests) of the total counts of the Maroon pigeon can be computed as $u_{11} = N_{11}/N_1 = 15/16 = 0.938$, whereas the proportion of habitat 2 (secondary forests) in the distribution of that species is $u_{21} = 1/16 = 0.062$.

We can now compute HUDS based on the Shannon index for *Columba thomensis* as (see Table 7.3)

$$\text{HUDS} = \exp\{ - [(15/16) * \ln (15/16) + (1/16) * \ln (1/16)] \}$$
$$= \exp\{ - [0.938 * \ln (0.938) + 0.063 * \ln (0.063)] \} = 1.26$$

Taking into account the fact that all the species have the same available habitat, the values of HUDS indicate that the maroon pigeon is the most specialized, occurring in two habitats, but in fact the distribution is almost equivalent to just one habitat (HUDS = 1.26). On the contrary, the bronze-naped pigeon occurs in the four habitats available with an equivalent number (HUDS) of 3.01 habitats, indicating that it is the more generalist of the pigeons of the island.

Similarly to what was done to measure the habitat use diversity of a species, we can measure the global habitat use diversity (GHUD). The first measure of the habitat use

Table 7.3 Results of the evaluation of habitat use diversity measured as the equivalent number of habitats based on the Shannon index for the different species and for all of the pigeons (we show the proportion of species j in habitat k of the total number of individuals of species j).

Bird species	Types of habitat				Total number of individuals counted (N_j)	Number of equivalent habitats (HUDS$_j$)
	Old-growth forest	Secondary forest	Shade forest	Non-forested		
Maroon pigeon	15 (0.938)	1 (0.063)	0 (0.000)	0 (0.000)	16	1.26
Lemon dove	90 (0.231)	220 (0.566)	71 (0.183)	8 (0.021)	389	2.86
Green pigeon	69 (0.479)	59 (0.410)	16 (0.111)	0 (0.000)	144	2.62
Bronze-naped pigeon	31 (0.048)	355 (0.552)	163 (0.253)	94 (0.146)	643	3.01
Total number of individuals counted ($N_{.k}$)	205 (0.172)	635 (0.533)	250 (0.210)	102 (0.086)	1192	GHUD = 3.24

diversity is the total number of habitats in the landscape that are used by all the species considered. However, similarly to what was done for the habitat use diversity of species, we can use the modified Shannon index to compute GHUD as

$$GHUD = \exp\left\{ -\sum [u_k \ln u_k] \right\}$$

where u_k is now the proportion of the individuals of all the species that used that habitat.

In the example above the global habitat use diversity was evaluated as 3.24 equivalent habitats. This value is higher than the values for the single species, indicating that different species use different habitats. It should be noted, however, that these calculations are done by adding the number of individuals of the different species that therefore are considered of equal importance.

7.4 The Relationship Between the Species Diversity of a Landscape and the Habitat Use Diversity of the Species

In the above sections we demonstrated that there are different measurements of diversity (expressed in the Hill series) and different types of diversity, those associated with species diversity (of habitats SDH and landscapes SDL) and those associated with habitat use diversity (of individual species HUDS and for the group of species considered GHUD).

In this section we clarify the relationships between these different types of diversity. In summary, the diagram of Figure 7.8 illustrates the issue. If diversity is measured as richness the equation illustrated in Figure 7.9 applies. In fact, if all species are generalists or use all habitats in the same proportions, we have GHUD = AHUDS and therefore the species diversity of the landscape will be equal to the average species diversity of its

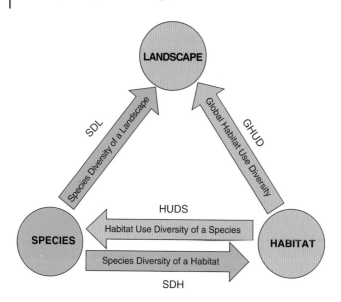

Figure 7.8 This diagram illustrates the relationships between the different types of diversity, showing that species diversity can apply to single habitats (SDH) or to whole landscapes (SDL), and that habitat use diversity can apply to individual species (HUDS) or to all the species globally (GHUD).

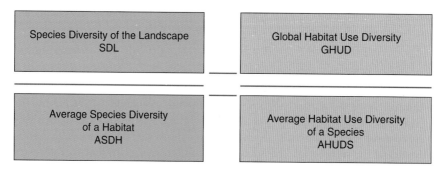

Figure 7.9 This diagram illustrates the equation relating the different types of diversity when measured as richness (SDL/ASDH = GHUD/AHUDS). From this equation it is easy to see that the total number of species (SDL) of a landscape composed of different habitats increases directly with the average species diversity of their habitats (ASDH) and with the number of habitats used by all the species (GHUD), but inversely with the average number of habitats used per species (AHUDS).

habitats (SDL = ASDH). On the contrary, if we have only very specialized species occurring only in one habitat (AHUDS = 1) then SDL = ASDH $*$ GHUD, meaning that the global habitat use diversity is extremely important in determining the species diversity of the landscape.

This type of analysis of species diversity components has been used in several studies[22]. Using richness, the species diversity of a landscape (SDL or *gamma* diversity) is decomposed into the average species diversity of habitats (ASDH or *alpha* diversity) and *beta* diversity, computed as the ratio global habitat use diversity/average habitat use diversity of a species (GHUD/AHUDS).

Figure 7.10 The American kestrel (*Falco sparverius*) in the collection of images of the birders of St Kitts and Nevis (top left). Views from Trinidad (top right) and St Kitts (below). *Sources:* American_Kestral_Oct_2012, Turtle_Beach_Pond_Area,_(M._Ryan), http://birds.impossiblework.com/bird-photos/(top left); http://mentalfloss.com/article/28673/why-do-they-call-it-trinidad-and-tobago (top right); http://media-cdn.tripadvisor.com/media/photo-s/01/b9/71/00/a-view-of-st-kitts-and.jpg (below).

A good example of the use of this decomposition of diversity is found in the work of Schluter and Ricklefs[23] with bird diversity in the Caribbean islands of Trinidad and St Kitts (Figure 7.10). The total species richness of the islands can be decomposed into the average species richness of their habitats (ASDH), the number of habitats used (GHUD), and the average number of habitats used by the species (AHUDS)[24]. Recall that SDL = ASDH $*$ GHUD/AHUS. The results are shown in Table 7.4.

Table 7.4 The decomposition of bird species diversity (SDL is here measured as the number of species) of the islands of Trinidad and St Kitts.

Island	Total number of species (SDL)	Average number of species per habitat (ASDH)	Number of habitats used in the island (GHUD)	Average number of habitats used per species (AHUDS)
Trinidad	108	28.2	9	2.33
St Kitts	20	11.9	9	5.26

Adapted from: Schluter, D. and Ricklefs, R.E. (1993) Species diversity. An introduction to the problems, in *Species Diversity in Ecological Communities. Historical and Geographical Perspectives* (eds R.E. Ricklefs and D. Schluter), The University of Chicago Press, Chicago, IL.

From the above example it is clear that the higher species richness of Trinidad is not a function of the number of habitats in the islands (which is equal) but of the higher species richness of the habitats on that island (28.2 in Trinidad compared to 11.9 in St Kitts) and also the more specialized bird species in Trinidad (2.33 habitats used per species) than in St Kitts (5.26 habitats used per species). This example illustrates the differences and the relationships between habitat use diversity and species diversity of landscapes.

Equivalent analyses may be done using approximations with indices that incorporate the abundance of species (as counts of individuals) instead of only presence–absence.

Using our example of the pigeons in São Tomé (H_1 from Table 7.2 and Figure 7.11), we can compute the average species diversity of its habitats (ASDH) as

$$ASDH = (3.34 + 2.52 + 2.25 + 1.32)/4 = 2.36$$

These results indicate that the species diversity in the island (SDL = 2.75) is only slightly higher than the average species diversity of its habitats (ASDH = 2.36) and lower than the species diversity of one of its habitats (old-growth forests – SDH = 3.34). In fact, all species are well represented in old-growth forests.

Figure 7.11 A view of the Atlantic island of São Tomé. *Source:* Rui Almeida, originally posted to Flickr as São Tomé – Ilhéu das Rolas HDR, CC BY 2.0, https://commons.wikimedia.org/w/index.php?curid=5629766.

Based on the results of Table 7.3, we can also compute the average habitat use diversity of all the species present (AHUD):

$$AHUD = (1.26 + 2.86 + 2.62 + 3.01)/4 = 2.44$$

This also results in that the global habitat use diversity (GHUD = 3.24) is not much higher than the average habitat use diversity of its species (AHUDS = 2.44), confirming some overlap in species composition of the various habitats.

However, when using this decomposition of species diversity of a landscape careful attention should be paid to which levels are considered and on their interrelationships, as it is clear that the different hierarchical levels contain different information[25].

7.5   Habitat Selection

Different species use landscapes in different ways. The different habitats present in the landscape provide food, shelter and breeding areas for the various species, but individual species differ in their ecological requirements and therefore use the landscape differently (Figure 7.12).

The analysis of habitat selection is therefore a common and important aspect of wildlife science[26] and a critical aspect for the conservation and management of a wildlife species. When species use resources (food items or habitats) disproportionately to their availability, the use is said to be selective, and many analytic procedures have been devised to compare the usage of these resources in comparison to their availability to the individual or to a population of a certain species.

It is important, however, to recognize the hierarchical nature of selection. A natural order of selection processes can be identified from the selection of the geographical range of a species to the selection of the home range of an individual or population within that range to the selection of types of habitat within the home range, or the selection of food items within the feeding site[27]. Many different study designs and tests have also been used to compare resource use and availability to individuals or populations[28].

Figure 7.12 James Peek, one of the authors of a pioneer study of habitat selection of moose (*Alces alces*) in Minnesota. *Sources:* Courtesy of James Peek (left); http://images2.citypages.com/imager/u/original/6549110/moose.jpg (right).

Table 7.5 Results of a study indicating the counts of occurrences (number observed) of moose tracks in different types of habitat (k) and showing the calculations comparing usage and availability to compute the selection ratio.

Habitat (k)	Number observed (ON_k)	Proportion used ($u_k = ON_k/N$)	Proportional availability in the landscape (p_k)	Number expected ($EN_k = p_kN$)	Selection ratio ($SR_k = ON_k/EN_k = u_k/p_k$)
Inside the burn, far from the edge	25	0.21	0.34	39.8	0.63
Inside the burn, close to the edge	22	0.19	0.10	11.7	1.88
Out of burn, close to the edge	30	0.26	0.10	11.7	2.56
Out of burn, far from the edge	40	0.34	0.46	53.8	0.74
Total	$N = 117$	1.00	1.00	117	

Adapted from Neu, C.W., Byers, C.R., and Peek, J.M. (1974) A technique for analysis of utilization–availability data. *Journal of Wildlife Management*, 38, 541–545.

In this section we will focus on the selection of habitats by individuals of a certain species within a certain landscape (defined as a mosaic of habitats). Habitat selection will be evaluated by the comparison of the relative use of the habitats to their availability in the landscape. In this example we will consider one of the first studies of habitat selection applied to the habitat use of moose in the winter of 1971–1972 after wildfires in northeastern Minnesota[29].

This example of habitat selection by moose after wildfires was provided in a study published in 1974[30] that might be considered as one of the earlier efforts to develop more quantitative and objective means of analyzing habitat use information[31]. The study observed the occurrence of moose tracks in different habitats and compared the observed numbers with the expected values as if moose occurred in each habitat in exact proportion to availability (Table 7.5).

The ratio between the observed number of occurrences in habitat k (ON_k) and the corresponding expected values (EN_k) gives an indication of the habitat selection. This ratio can also be computed using the observed proportional usage of the habitat ($u_k = ON_k/N$) and proportional availability of that habitat in the landscape (p_k), where N is the total number of animals, in this case moose tracks, observed.

This selection ratio has been used extensively since 1938–1939 after being applied in pioneer studies on duck nesting[32] and on food preferences of trout[33], taking the initial name of forage ratio. We will use the broader designation of the selection ratio (SR):

$$SR_k = ON_k/EN_k = u_k/p_k$$

This index varies from zero (no usage) to infinity (usage infinitely larger than availability), taking the value of one when there is no selectivity, that is, when the habitat is used in proportion to its availability in the landscape.

From the results of the calculations it is possible to conclude that moose were selecting areas close to the edge of the burn (inside or outside) where SR > 1, meaning that in those habitats proportional usage was higher than proportional availability and the numbers observed were therefore higher than those expected if there was no selectivity.

Other authors prefer an index ranging from −1 (no usage) to +1 (usage infinitely larger than availability), going through zero ($u_k = p_k$) when there is no selectivity. Under this concept, Ivlev developed, in studies on the feeding ecology of fishes published in 1961[34], his widely used index of electivity (IE):

$$IE_k = (ON_k - EN_k)/(ON_k + EN_k) = (u_k - p_k)/(u_k + p_k)$$

For example, the index of selectivity of moose in the interior of the burned area would be $IE_1 = (25 - 39.8)/(25 + 39.8) = -0.23$, indicating negative selection while still inside the burn, but close to the edge the electivity index would be $IE_2 = (22 - 11.7)/(22 + 11.7) = +0.29$, indicating positive selection.

Many statistical tests of different complexity were proposed for these studies. However, when analyzing counts of individuals or signs in different habitats a simple statistical test can be used by calculating a chi-square (χ^2) statistic using observed (ON_k) and expected (EN_k) values as

$$\chi^2_{calc} = \sum [(ON_k - EN_k)^2/EN_k)$$

$$\chi^2_{calc} = (25 - 39.8)^2/39.8 + (22 - 11.7)^2/11.7 + (30 - 11.7)^2/11.7$$

$$+ (40 - 53.8)^2/53.8 = 46.73$$

For the statistical test the calculated value of chi-square has to be compared with the corresponding value found for the chosen level of significance (typically 0.05) and the number of degrees of freedom (the number of classes k minus one, in this case $4 - 1 = 3$). The value of chi-square found in the statistical tables for the 0.05 level of significance and 3 degrees of freedom ($\chi^2_{tab.\ 0.05,\ 3d.f.}$) is 7.815. As our calculated value is higher ($\chi^2_{calc} > \chi^2_{tab}$) we may conclude that our statistical test supports the conclusion that the observations are not distributed according to proportions of the different habitats in the landscape. Habitat selection does occur.

Many other types of analyses are available for the comparison of habitat use and availability. Between those, it is worth mentioning the studies based on compositional analyses following the works of Aebisher and others in 1993[35]. The method of compositional analysis has since been applied in many situations, such as those of badgers in Finland[36]. However, statistical analyses of use/availability data are always necessary to test the null hypothesis that habitats were used in proportion to their availability. Also, many different equivalent approaches have been taken, such as that of comparing habitats of observed elk locations with those of random points[37].

Habitat selection analysis methods described in this chapter are not limited to animal habitats, but also apply to plants or plant communities. Habitat selection for plant species is commonly referred to as environmental niche models, defining envelopes of environmental variables where a species can be sustained. Habitat variables commonly include precipitation, temperature, topography, soil, aspect, and elevation (Figure 7.13).

Similar habitat selection analyses can be performed for landscape processes instead of organisms, and these landscape processes can be associated with flows of materials or disturbances. Analyses can be performed using point locations or home ranges of animals, with location of points, lines, or polygons representing disturbances.

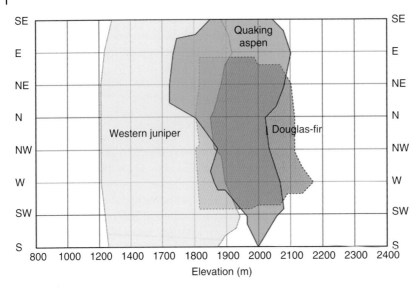

Figure 7.13 Habitat selection for quaking aspen (*Populus tremuloides*), western juniper (*Juniperus occidentalis* spp. *occidentalis*), and Douglas-fir (*Pseudotsuga menziesii*) are here illustrated with regards to elevation and aspect in the Owyhee Mountains in western United States. In this region, the preferred habitat for quaking aspen overlaps that of western juniper and Douglas-fir. However, quaking aspen is the only tree species occupying southern and southeastern aspects at elevations above 1900 m where competition from the other species is limited due to environmental conditions[38]. Quaking aspen in southwestern Idaho (right). Photo by Eva K. Strand.

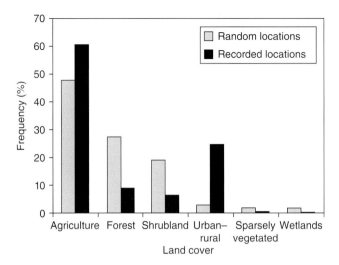

Figure 7.14 Comparison between the frequency of random locations and locations where fire ignitions were recorded in Portugal showing that ignitions were mostly "selecting" areas in the urban–rural interface and, to a minor extent, agricultural areas. *Source:* Catry, F.X., Rego, F.C., Bação, F., and Moreira, F. (2009) Modeling and mapping wildfire ignition risk in Portugal. *International Journal of Wildland Fire*, 18, 921–931.

Using fire as an example we can use the locations of fire ignitions as points, the fire perimeters as lines, or the area burned as patches. An example of the use of this type of analysis in relation to fire can be provided by the comparison between the land cover types of observed locations of fire ignitions in Portugal with those of points located randomly in the country[39] (Figure 7.14).

Regardless of the organism or process under analysis, the study designs and the corresponding statistical tests have to be carefully selected. Good summaries of study designs and tests for comparing habitat use and availability were provided by Thomas and Taylor in 1990[40] and in 2006[41] in their review of 87 articles published in the *Journal of Wildlife Management*. If there are significant differences between habitat use and availability for a species we can investigate its degree of specialization.

An interesting measure of specialization that may use u_k and p_k was developed in 1967 by Theil[42] in the context of the use of information theory in economics. The Theil index (THI) is calculated as follows:

$$\text{THI} = (1/m) \sum \left[(u_k/p_k) \ln (u_k/p_k) \right]$$

where m is the number of classes (habitats) in the landscape and u_k and p_k are, respectively, the proportional use and availability of habitat k.

As u_i/p_i is the selection ratio (SR) presented earlier in this text, we could also compute the Theil index as

$$\text{THI} = (1/m) \sum \left[(\text{SR}_k) \ln (\text{SR}_k) \right]$$

This index takes the value of zero when there is no selection ($u_k = p_k$ or all $\text{SR}_k = 1$) and differs increasingly from zero with increasing specialization. Notice that zero is not the

Figure 7.15 The maroon pigeon (*Columba thomensis*) found in São Tomé e Príncipe (left), an image of the vegetation of the Island (center), and Mariana Carvalho (right), a researcher on the ecology of São Tomé pigeons. *Sources:* Photos courtesy of Mariana Carvalho (left and center); Courtesy of Ricardo Rocha (right).

lower bound of this index. The Theil index is positive when the positive selections are more influential than the negative ones and turns negative the other way around.

We can turn once more to our earlier example of the pigeons in São Tomé as an example of habitat selection (Figure 7.15 and Table 7.6).

The selection ratio is simply calculated by dividing u_{kj} by p_k, as illustrated next. For example, the selection of old-growth forest by the maroon pigeon is $SR_{11} = u_{11}/p_1 = 0.938/0.283 = 3.31$, indicating a very positive selection since the value of SR is well above unity. Figure 7.16 displays the values of the Selection Ratios for all species.

Table 7.6 The proportions of habitat use (u_{kj}) for each species j (derived from the data in Table 7.1) and the proportion of habitat k in the sampled landscape (p_k).

		Types of habitat			
Bird species	Common name English (and Portuguese)	Old-growth forest	Secondary forest	Shade forest	Non-forested
Columba thomensis	Maroon pigeon (Pombo-do-mato)	0.938	0.062	0.000	0.000
Columba simplex	Lemon dove (Muncanha)	0.231	0.566	0.182	0.021
Treron sanctithomae	São Tomé green pigeon (Céssia)	0.479	0.410	0.111	0.000
Columba malherbii	Bronze-naped pigeon (Rola)	0.048	0.552	0.254	0.146
Proportion of habitat in the sampled landscape (p_k)		0.283	0.494	0.152	0.071

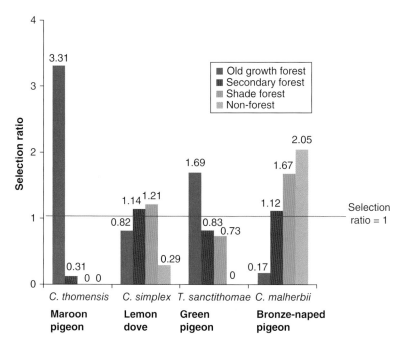

Figure 7.16 The values of the selection ratio for the different bird species and the different types of habitat based on the study and the data presented in Table 7.5.

From Figure 7.16 it is apparent that the different species show different patterns in the utilization of the habitats. Whereas two species (the maroon pigeon and the green pigeon) show very strong positive selection of old-growth forests, another species (the bronze-naped pigeon) shows almost the opposite behavior. The lemon dove is intermediate between the first two extreme situations.

Finally, we can compute the chi-square statistics to test the null hypothesis that the distribution of the birds between habitats is the same as the proportion of those habitats in the landscape, that is, if all $u_{kj} = p_k$.

We can calculate the expected number of counts of the maroon pigeon in old-growth forests. If there was no selectivity, EN_{11} would be equal to the proportion of old-growth forests in the landscape ($p_1 = 0.283$) multiplied by the total number of counts of that species ($N_1 = 16$), yielding the value 4.53 individuals. As the observed count (ON_{ii}) was 15 we can compute the ratio ($ON_{11}/EN_{11} = 3.31$) confirming the high value of the selection ratio shown in the graph.

If we calculate the other expected counts for all the species we can calculate the chi-square statistic as $\chi^2_{calc} = \sum[(ON_k - EN_k)^2/EN_k)$, with the following results:

For the maroon pigeon:

$$\chi^2 = (15 - 4.53)^2/4.53 + (1 - 7.91)^2/7.91$$
$$+ (0 - 2.42)^2/2.42 + (0 - 1.14)^2/1.14 = 33.84$$

For the lemon dove:

$$\chi^2 = (90 - 110.03)^2/110.03 + (220 - 192.28)^2/192.28$$
$$+ (71 - 58.91)^2/58.91 + (8 - 27.79)^2/27.79 = 24.22$$

For the green pigeon:

$$\chi^2 = (69 - 40.73)^2/40.73 + (59 - 71.18)^2/71.18$$
$$+ (16 - 21.81)^2/21.81 + (0 - 10.29)^2/10.29 = 33.53$$

For the bronze naped pigeon:

$$\chi^2 = (31 - 181.88)^2/181.88 + (355 - 317.83)^2/317.83$$
$$+ (163 - 97.37)^2/97.37 + (94 - 45.93)^2/45.93 = 224.06$$

The value of chi-square found in any statistical table for the 0.05 level of significance and 3 degrees of freedom (the number of habitats minus one) is 7.82. Since our calculated values are well above that threshold, we may conclude that our statistical tests support the conclusion that the pigeons are not distributed in according to the proportions of the different habitats in the landscape. Habitat selection is occurring.

We may also compute the Theil index for the different species in order to understand their degree of specialization. Using the equation based on the already computed selection ratios:

$$THI = (1/m) \sum [(SR_k) \ln (SR_k)]$$

we have for all the species:

For the maroon pigeon:

$$THI = (1/4) * [3.31 \ln (3.31) + 0.13 \ln (0.13) + 0 + 0]$$
$$= 3.71/4 = 0.93$$

For the lemon dove:

$$THI = (1/4) * [0.82 \ln (0.82) + 1.14 \ln (1.14) + 1.21 \ln (1.21) + 0.29 \ln (0.29)]$$
$$= -0.14/4 = -0.04$$

For the green pigeon:

$$THI = (1/4) * [1.69 \ln (1.69) + 0.83 \ln (0.83) + 0.73 \ln (0.73) + 0]$$
$$= 0.51/4 = 0.13$$

For the bronze-naped pigeon:

$$THI = (1/4) * [0.17 \ln (0.17) + 1.12 \ln (1.12) + 1.67 \ln (1.67) + 2.05 \ln (2.05)]$$
$$= 2.15/4 = 0.54$$

These values also confirm that the maroon pigeon is the more specialized pigeon as it selects old-growth forests almost exclusively. The Theil index also confirms the bronze-naped pigeon as the second most selective pigeon for the exact opposite reasons to the first. The other two species (the lemon dove and the green pigeon) show much less selectivity.

It should be noted that the chi-square statistic is much higher for the bronze-naped pigeon whereas the Theil index is higher for the maroon pigeon. This is due to the fact that the bronze-naped pigeon has many more counts than the maroon pigeon, therefore allowing for more statistical evidence of the differences.

In conclusion, this example clearly demonstrates the importance of conserving the old-growth forests to maintain the diversity of the pigeons in the island of São Tomé. This is due to the fact that an old-growth forest is the habitat with the highest species diversity (higher diversity than the global landscape) but also because there are species such as the maroon pigeon that are very specialized for this habitat and are therefore very affected by possible habitat losses.

7.6 Landscape Composition and Diversity

We saw in the previous sections how diversity could be measured by the different indices of the Hill family, how these measurements were used to assess species diversity of habitats and landscapes, how the species diversity of landscapes was related to the habitat use diversity of the species, and how species select habitats within the landscape. In this section we demonstrate how landscape composition and its habitat diversity are related to its species diversity.

It is important to note that the first uses of diversity indices in ecology focused on species diversity but, as landscape ecology developed, diversity indices have also been used extensively to analyze changes in the diversity of landscape composition[43,44,45]. In these cases, habitat class is used in place of a species and the area of a landscape occupied by each patch class is often used as the measure of abundance.

Using the modified Shannon index H_1 we can calculate the habitat diversity of the landscape (HDL) using the proportions p_k of the different k habitats as

$$HDL = \exp\left\{ -\sum [p_k \ln (p_k)] \right\}$$

We can use again the example of São Tomé to illustrate these calculations.

Using the proportions of the different habitats in the sampled landscape (p_i) that have already been presented in Table 7.6, we can compute the habitat diversity of the landscape:

$$HDL = \exp \{ - [0.283 * \ln (0.283) + 0.494 * \ln (0.494) + 0.152 * \ln (0.152)$$
$$+ 0.071 * \ln (0.071)\} = 3.25$$

This value is not very different from the global habitat use diversity of that landscape (GHUD = 3.24). However, we saw that different species used habitats differently and that a selection ratio (SR_k) could be computed to compare the proportion of habitat used by the species (u_k) with the proportion of that habitat available in the landscape (p_k):

$$SR_k = u_k / p_k$$

We can therefore write that

$$u_k = p_k SR_k$$

Therefore the habitat use diversity of a species (or a group of species) can be written as

$$\text{HUDS} = \exp\left\{ -\sum [(p_k \text{SR}_k)/\ln(p_k \text{SR}_k)] \right\}$$

This equation indicates that with generalist species, where the selection ratios are close to unity, the habitat use diversity of that species is close to the habitat diversity of the landscape. The above equation is simply a weighted version of the modified Shannon index where the weights are the selection ratios.

However, this equation can be applied to any situation where the different habitats are valued differently. These values could be associated with their preference by animal or plant species or by their economic or conservation value, and therefore different weights (w_i) could be assigned to the different habitats according to the objective to yield a value for the weighted habitat diversity of the landscape (WHDL):

$$\text{WHDL} = \exp\left\{ -\sum [(w_k p_k)/\ln(w_k p_k)] \right\}$$

These types of weighted diversity indices and their mathematical developments based on the Shannon and Simpson indices have been studied and proposed with many possible biological and economical applications relative to the assessment of diversity measured by compositional proportions of a system such as a landscape[46,47].

Due to their simplicity and power, simple diversity indices have been used extensively since the pioneer works of Shannon and Simpson in both community and landscape ecology. However, they have also been criticized in the ecological literature because they were not conveying any information about individual species or habitats. The use of weighted diversity indices as a response to those criticisms is now very promising.

Key Points

- Biodiversity is defined as the variety of life on Earth, including all organisms, species, and populations, and the genetic variation among these, and their complex assemblages of communities and ecosystems. It also refers to the interrelatedness of genes, species, and ecosystems and their interactions with the environment.
- Species richness is the first described index of diversity. It is the total number of species present within the study area. Because a count of all species is virtually impossible, species richness normally includes a small portion of the total biota, for example vascular plants or breeding birds.
- Species diversity indices account for the abundance of species. An area where all species are of equal abundance is considered more diverse compared to an area with the same number of species but where some species are common and others rare.
- Evenness indices are also a measure of diversity but they are adjusted for the number of species present, the species richness.
- Richness, diversity, and evenness indices can also be applied to landscapes, where the number of patch types represents patch richness and diversity and evenness represent the diversity and evenness of area distribution between patch types.
- Different species use landscapes in different ways; for example, habitats provide food, shelter and breeding areas. The analysis of habitat selection attempts to identify habitats that are used disproportionally to their availability in the landscape, which is important information for the conservation and management of a species.

Endnotes

1 http://www.unep.org/wed/2010/english/PDF/BIODIVERSITY_FACTSHEET.pdf.

2 Magurran, A. (1988) *Ecological Diversity and Its Measurement*, Springer, The Netherlands.

3 Fisher, R.A, Corbett, A.S., and Williams, C.B. (1943) The relation between the number of species and the number of individuals in a random sample of an animal population. *Journal of Animal Ecology*, 12, 42–58.

4 Heip, C., Herman, P., and Soetaert, K. (1998) Indices of diversity and evenness. *Oceanis*, 24, 61–87.

5 Shannon, C. (1948) A mathematical theory of communication. *Bell System Technical Journal*, 27, 379–423, 623–656.

6 Shannon, C. and Weaver, W. (1949) *The Mathematical Theory of Communication*, University of Illinois Press, Urbana, IL.

7 Simpson, E.H. (1949) Measurement of diversity. *Nature*, 163, 688.

8 Berger, W.H. and Parker, F.L. (1970) Diversity of planktonic Foraminifera in deep-sea sediments. *Science*, 168, 1345–1347.

9 Shannon, C. (1948) A mathematical theory of communication. *Bell System Technical Journal*, 27, 379–423, 623–656.

10 Weaver, W. and Shannon, C. (1963) *The Mathematical Theory of Communication*, University of Illinois Press, Urbana, IL.

11 Simpson, E.H. (1949) Measurement of diversity. *Nature*, 163, 688.

12 Berger, W.H. and Parker, F.L. (1970) Diversity of planktonic Foraminifera in deep-sea sediments. *Science*, 168, 1345–1347.

13 Whittaker, R.H. (1965) Dominance and diversity in land plant communities. *Science*, 147, 250–260.

14 Magurran, A. (1988) *Ecological Diversity and Its Measurement*, Springer, The Netherlands.

15 Hill, M.O. (1973) Diversity and evenness: a unifying notation and its consequences. *Ecology*, 54, 427–432.

16 Heip, C., Herman, P., and Soetaert, K. (1998) Indices of diversity and evenness. *Oceanis*, 24, 61–87.

17 Magurran, A. (1988) *Ecological Diversity and Its Measurement*, Springer, The Netherlands.

18 Whittaker, R.H. (1972) Evolution and measurement of species diversity. *Taxon*, 21, 213–251.

19 Whittaker, R.H. (1977) Evolution of species diversity in land communities. *Evolutionary Biology*, 10, 1–67.

20 MacArthur, R. (1965) Patterns of species diversity. *Biological Reviews*, 40, 510–533.

21 Carvalho, M., Fa, J.E., Rego, F.C., *et al.* (2015) Factors influencing the distribution and abundance of endemic pigeons in São Tomé Island (Gulf of Guinea). *Bird Conservation International*, 25, 71–86.

22 Routledge, R.D. (1977) On Whittaker's components of diversity. *Ecology*, 58, 1120–1127.

23 Schluter, D. and Ricklefs, R.E. (1993) Species diversity. An introduction to the problems, in *Species Diversity in Ecological Communities. Historical and Geographical Perspectives* (eds R.E. Ricklefs and D. Schluter), The University of Chicago Press, Chicago, IL.

24 Schluter, D. and Ricklefs, R.E. (1993) Species diversity. An introduction to the problems, in *Species Diversity in Ecological Communities. Historical and Geographical Perspectives* (eds R.E. Ricklefs and D. Schluter), The University of Chicago Press, Chicago, IL.

25 Weins, J.A. (1989) *The Ecology of Bird Communities*, Cambridge University Press, Cambridge, MA.

26 Alldredge, J.R. and Ratti, J.T. (1986) Comparison of some statistical techniques for analysis of resource selection. *Journal of Wildlife Management*, 50, 157–165.

27 Johnson, D.H. (1980) The comparison of usage and availability measurements for evaluating resource preference. *Ecology*, 61, 65–71.

28 Thomas, D.L. and Taylor, E.J. (1990) Study designs and tests for comparing resource use and availability. *Journal of Wildlife Management*, 54, 322–330.

29 Neu, C.W., Byers, C.R., and Peek, J.M. (1974) A technique for analysis of utilization–availability data. *Journal of Wildlife Management*, 38, 541–545.

30 Neu, C.W., Byers, C.R., and Peek, J.M. (1974) A technique for analysis of utilization–availability data. *Journal of Wildlife Management*, 38, 541–545.

31 http://garfield.library.upenn.edu/classics1992/A1992HT79800001.pdf.

32 Williams, C.S. and Marshall, W.H. (1938) Duck nesting studies, Bear River Migratory Bird Refuge, Utah, 1937. *Journal of Wildlife Management*, 2, 29–48.

33 Hess, A.D. and Rainwater, J.H. (1939) A method for measuring the food preference of trout. *Copeia*, 3, 154–157.

34 Ivlev, V.S. (1961) *Experimental Ecology of the Feeding of Fishes*, Yale University Press, New Haven, CN.

35 Aebisher, N.J., Robertson, P.A., and Kenward, R.E. (1993) Compositional analysis of habitat use from animal radio-tracking data. *Ecology*, 74, 1313–1325.

36 Kauhala, K. and Auttila, M. (2010) Estimating habitat selection of badgers – a test between different methods. *Folia Zoologica*, 59, 16–25.

37 Marcum, C.L. and Loftsgaarden, D.O. (1980) A non-mapping technique for studying habitat preferences. *Journal of Wildlife Management*, 44, 963–968.

38 Strand E.K., Vierling, L.A, Bunting, S.C., and Gessler, P.E. (2009) Quantifying successional rates in western aspen woodlands: Current conditions, future predictions. *Forest Ecology and Management*, 257, 1705–1715.

39 Catry, F.X., Rego, F.C., Bação, F., and Moreira, F. (2009) Modeling and mapping wildfire ignition risk in Portugal. *International Journal of Wildland Fire*, 18, 921–931.

40 Thomas, D.L. and Taylor, E.J. (1990) Study designs and tests for comparing resource use and availability. *Journal of Wildlife Management*, 54, 322–330.

41 Thomas, D.L., and Taylor, E.J. (2006) Study designs and tests for comparing resource use and availability II. *Journal of Wildlife Management*, 70, 324–336.

42 Theil, H. (1967) *Economics and Information Theory*, Rand McNally and Company, Chicago, IL.

43 O'Neill, R.V., Krummel, J.R., Gardner, R.H., *et al.* (1988) Indices of landscape pattern. *Landscape Ecology*, 1, 153–162.

44 Romme, W.H. (1982) Fire and landscape diversity in subalpine forests of Yellowstone National Park. *Ecological Monographs*, 52, 199–221.

45 Turner, M.G. (1990) Spatial and temporal analysis of landscape pattern. *Landscape Ecology*, 4, 21–30.

46 Casquilho, J., Neves, M., and Rego, F. (1997) Extensões da função de Shannon e equilíbrio de proporções – uma aplicação ao mosaico de paisagem. *Anais Instituto Superior Agronomia*, 46, 77–99.

47 Casquilho, J.P. (2016) A methodology to determine the maximum value of weighted Gini–Simpson index. *SpringerPlus*, 5, 1143.

8

Landscape Pattern: Composition and Configuration

Quantifying landscape pattern is essential to understand landscape processes. Landscape composition is a fundamental characteristic, the importance of which was discussed in the previous chapter. However, landscape pattern is not fully described by composition. Configuration is another important characteristic that complements the description of pattern. In this chapter we discuss how composition and configuration represent different aspects of pattern and how configuration is measured and interpreted.

8.1 Composition and Configuration Represent Different Aspects of Landscapes

The number and the proportion of different classes in the landscape determine its composition. The classes can be established in a number of different ways based on different factors such as land use, land ownership, or vegetation cover type. One of the most common ways to express landscape composition, however, is with respect to the types of habitat present. Generally we will use this characteristic to express landscape composition in this chapter.

The composition of a landscape is generally measured by different diversity indices, from the simplest, landscape richness (the number of habitats present in the landscape), to more complex indices that take into account the proportions (p_k) of the different k habitats.

The proportions of the different habitats in the landscape can be evaluated by any attribute as the number of individuals (as seen in the previous section), the biomass, or the area occupied. Here, we will assume that landscape composition is evaluated by the proportion of the area occupied by the different habitats, and we will quantify the diversity of landscape composition by the habitat diversity of the landscape (HDL) using the modified Shannon index as

$$HDL = \exp \sum [-(p_k) \ln (p_k)]$$

where p_k is the proportion of the habitat k in the landscape.

Applied Landscape Ecology, First Edition. Francisco Castro Rego, Stephen C. Bunting, Eva Kristina Strand and Paulo Godinho-Ferreira.
© 2019 John Wiley & Sons Ltd. Published 2019 by John Wiley & Sons Ltd.

However, the measurement of landscape composition and its diversity is generally not enough to adequately represent the functions of the landscape, the flow of organisms, water, and nutrients. The spatial configuration of the patches of the different habitats in the landscape is equally important. Landscape composition therefore needs to be complemented with measures of landscape configuration.

In practical terms, as habitat diversity of a landscape (HDL) applies to measure the composition of that landscape by using the proportions of area occupied by the different habitats, it is possible to consider, in raster representations, the proportions of all types of cell adjacencies, including those inside a patch, to measure the configuration of the same landscape.

In this chapter we will use the example in Figure 8.1 of raster representations of two landscapes with three habitats with the same proportions, that is, the same area (25 cells), and therefore the same composition (equal proportion of habitats) but different configurations.

The diversity of the composition of the landscapes represented in Figure 8.1 is measured as usual based on the proportions (p_k) of cells in the different classes or habitats. For both cases:

$$p_A = 13/25 = 0.52$$
$$p_B = 4/25 = 0.16$$
$$p_C = 8/25 = 0.32$$

and the habitat diversity of the landscapes (HDL) can be computed as\

$$\text{HDL} = \exp \sum \left[-(p_k) \ln (p_k) \right] = \exp - [0.52 * \ln (0.52) + 0.16 * \ln (0.16)$$
$$+ 0.32 * \ln (0.32)] = 2.72$$

This result indicates that the habitat diversity of the two landscapes is equal, and equivalent to 2.72 habitats. As the richness (the number of habitats in the landscape – m) is 3, the evenness (E_1) of the composition of the landscapes can be calculated using the Shannon index as

$$E_1 = \text{HDL}/m = 2.72/3 = 0.91$$

This is a relatively high number (close to unity), indicating that all the classes are nearly equally represented in the landscape. Evenness equals 1 for those landscapes where all classes are equally abundant in the landscape. The evenness value approaches zero as one class becomes dominant.

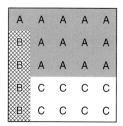
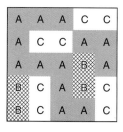

Figure 8.1 Two raster representations of landscapes with the same composition but different configurations. The landscape on the right shows a much more complex configuration than the much simpler landscape on the left.

As the two landscapes are obviously different but have the same composition and therefore the same habitat diversity, we need to consider other characteristics to differentiate the two landscapes. These characteristics are related to configuration.

Measurements of landscape configuration are important to reveal the pattern, or the structure, of the landscape, which is generally defined as the spatial relationship of the elements present in that landscape[1]. Landscape pattern or structure may be associated with different characteristics of the landscape and a number of different metrics have been proposed to measure it.

8.2 Configuration Assessed by Patch Numbers, Sizes, Perimeters, and Shapes

One of the simplest measures of landscape configuration is the mean patch area of the landscape (MPA) which can be computed easily for the whole landscape as:

$$\text{Landscape mean patch area (MPA)} = \text{total landscape area (TA)}/$$
$$\text{total number of patches (NP)}$$

or

$$\text{MPA} = \text{TA}/\text{NP}$$

The mean patch area of the landscape is the inverse of another common metric, patch density (λ_p), which is the total number of patches divided by the area of the landscape:

$$\lambda_p = \text{NP}/\text{TA}$$

From the observation of the two landscapes represented in Figure 8.1 it is possible to observe that the first landscape has only 3 patches whereas the second landscape, with a more complex configuration, has 8 patches. The mean patch area of the first landscape is therefore MPA = TA/NP = 25/3 = 8.3, whereas the value of MPA for the second landscape is MPA = TA/NP = 25/8 = 3.1, indicating a much more fragmented landscape. Obviously patch density shows the opposite results, with the first landscape having a patch density of λ_p = 3/25 = 0.12, a value much lower than that of the patch density of the second landscape: λ_p = 8/25 = 0.32. These two landscape metrics are very simple as they only require counting the total number of patches in the landscape.

However, a high number of landscape metrics have been developed to assess the configuration of the landscapes. From an analysis of 55 of those metrics, Riiters and others [2] concluded that the pattern may be adequately represented by as few as 6 univariate metrics, involving perimeter–area relationships, the number of classes, of patches and the adjacency of cells.

Starting with perimeter–area relationships and using the analogy with patches, we can now establish relationships between total edge and area for the whole landscape. Many configuration measures for the landscape compute the mean shape of all its patches, their mean patch fractal dimension, or mean core area index. It is also possible, by analogy, to use, for the landscape level, the total edge length (TL) and the total landscape area (TA) in a way similar to the patch perimeter and the patch area to derive a mean patch shape.

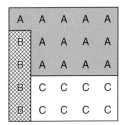 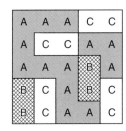

Figure 8.2 The same two landscapes represented in Figure 8.1 showing the edges between habitats. In the first landscape we have 9 edges whereas in the second landscape we have 23 edges.

The mean patch shape, as with many other shape indices, can be calculated at the individual class level and at the whole landscape level.

However, the simplest of all the measurements of the complexity of the configuration using perimeter–area relationships is the total edge length or edge line density (total edge length/total landscape area). We have calculated the edge line density for the same two landscapes, also represented in Figure 8.2.

We can now calculate the corresponding indices:

For the first landscape, the total edge line (TL) = 9 and the edge line density (λ_l) = TL/ TA = 9/25 = 0.36

For the second landscape, TL = 23 and we have the edge line density (λ_l) = TL/TA = 23/25 = 0.92

We can now say that the complexity of the configuration of the second landscape is much greater than that of the first landscape, as was easily observed.

Other measures of complexity based on the relationships between the total edge and landscape area have been proposed. That is the case for the landscape shape index (LSHAPE), which is equivalent for the whole landscape of the shape indices used to describe the complexity of single patches. For raster representations such as those in our example we have

$$\text{LSHAPE} = \text{TL} / \left[4(\text{TA})^{1/2} \right]$$

In our case, with total area TA = 25 for both examples, we can calculate the values for LSHAPE to be:

For the first landscape LSHAPE = 9/20 = 0.45
For the second landscape we have LSHAPE = 23/20 = 1.15

These calculations give exactly the same information as the edge line density metric.

Both of the indices presented can be applied for both raster and vector representations as they use only information on the perimeters of the patches (edges) and total area. For vector representations the landscape shape index (LSHAPE) is slightly modified:

$$\text{LSHAPE} = \text{TL} / \left[2(\pi \, \text{TA})^{1/2} \right]$$

As seen before, the landscape shape indices convey the same information as the edge line density, which is a much more readily interpretable measurement.

Average values of the shape indices of all patches of all types in the landscape (MSI) may also be applied and used as indicators of fragmentation of the landscape:

$$MSI = \sum\sum SHAPE_{ik}/NP$$

Many studies have been made in relating the complexity of landscape configuration with different ecological aspects, including species richness. Many of these studies use indices based on the number of patches, their size, perimeter, and shape.

These relationships can be illustrated by a study in the rural landscapes of eastern Austria that relates landscape patch shape complexity with species richness of vascular plants and bryophytes[3]. The authors found strong significant correlations between many patch area and shape indices and species richness for both vascular plants and bryophytes. Several factors could be partly responsible for the decreasing landscape complexity with increasing agriculture: (a) many of the agricultural features are rectangular in shape with abrupt boundaries, (b) increasing agriculture always reduces the more natural patch areas in the landscape, and (c) intensification of agriculture usually increases the size of agricultural patches (Figure 8.3).

These results indicate that landscape configuration indices are useful in relation to plant species diversity, but they also show that many of the indices give the same or very similar indications. The total numbers of patches, mean patch size, edge line density, mean patch edge, mean shape index, and other indices of shape complexity are indicators of fragmentation and related to species diversity (Table 8.1).

Figure 8.3 A rural landscape with vineyards in Austria. *Source:* https://commons.wikimedia.org/wiki/File:Terraced_vineyards_and_village_in_Wachau.jpg.

Table 8.1 Correlation coefficients between some indices reflecting landscape configuration and richness of vascular plants and bryophytes in the work in Austrian landscapes.

Index type	Index	Vascular	Bryophytes
Patch	Mean patch area (MPA)	−0.68[*]	−0.58[*]
Edge	Edge line density (λ_1)	0.56[*]	0.57[*]
Shape complexity	Mean shape index (MSI)	−0.70[*]	−0.51[*]

[*] Significant at $P < 0.01$.
Adapted from Moser, D., Zechmeister, H.G., Plutzar, C., et al. (2002) Landscape patch shape complexity as an effective measure for plant species richness in rural landscapes. *Landscape Ecology*, 17, 657–669.

8.3 Edge Contrast

The indices presented above are based on patch number, perimeters, areas, and shapes but they do not differentiate between types of edge lines. One of the possibilities of differentiating between different types of edges is to use the concept of edge contrast. The general approach is to assign a weight to a type of edge line based on some known characteristic differentiating the two patch types j and k. The contrast has been based on characteristics such as vertical structural complexity or successional stage[4].

The contrast weight (CON_{jk}) is intended to express the differences at the boundary between the two patch types j and k related to the functional aspects of the edge line (Figure 8.4).

We can consider the hypothesis that edge lines AB, BC, and AC have different values for contrast as they represent adjacencies between different habitats.

Using this approach we can compute an edge contrast index (EDCON) for the whole landscape as a weighted average of the contrasts for all the edge lines. This is calculated as the sum of the patch edge lengths between different patch types (L_{jk}) multiplied by their corresponding contrast weights (CON_{jk}), divided by the total edge length (TL = $\sum L_{jk}$):

$$EDCON = \sum \left(CON_{jk} L_{jk} \right) / TL$$

In the example shown in Figure 8.4, we can see that the edge length between A and B cells is $L_{AB} = 3$. Similarly, we have $L_{AC} = 4$ and $L_{BC} = 2$.

Let us suppose now that the contrast between A and B (both forest types, for example) is much smaller than the contrast between A and C and between B and C. For example:

Edge type	Edge length (L_{jk})	Contrast weight (CON_{jk})
AB	3	2
AC	4	5
BC	2	3

We would then have the edge contrast index (EDCON) for the first landscape computed as

$$EDCON = 2*3 + 5*4 + 3*2/9 = 32/9 = 3.56$$

Figure 8.4 Representation of the same first landscape as shown in Figures 8.1 and 8.2 showing different types of edge lines between adjacent cells.

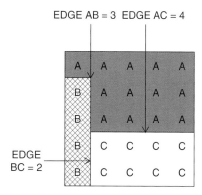

Similarly for the second landscape (Figure 8.2), where $L_{AB} = 6$, $L_{AC} = 14$, and $L_{BC} = 3$, we have

$$EDCON = (2*6 + 5*14 + 3*3)/23 = 91/23 = 3.96$$

This indicates that, given the weights used, edge lines are more contrasted in the second landscape (Figure 8.2) as a large proportion of the edges are between patch types A and C, which are those that have more contrast.

This concept of edge contrast has been used in some studies. In one of such studies on the effect of landscape structure on breeding birds in the Oregon Coast Range[5] (Figure 8.5), the authors selected 30 landscapes reflecting a range of disturbance (varying amounts of seral vegetation) with 27 patch types (22 forested, 5 nonforested) and sampled the birds at grid points in each landscape. The authors developed an edge contrast index and used principal components analysis to reduce the number of variables. A first component axis was defined as being heavily weighted by patch shape and edge contrast. Landscapes with high positive values in that axis contained late-seral forest vegetation that was distributed in patches with more complex shapes, greater edge density,

Figure 8.5 Forests of the western Oregon. Mount Hood is shown in the background. *Source:* Wikimedia Commons, https://commons.wikimedia.org/wiki/File:Mt_Hood_Natl_Forest.jpg.

Figure 8.6 The Hammond's flycatcher (*Empidonax hammondii*) (top left) and the western wood pewee (*Contopus sordidulus*) (top right). In the bottom the distribution (mean and 95% confidence intervals) of these two bird species of western Oregon is along a gradient (first principal component axis) of increased complexity and greater edge contrast. The Hammond's flycatcher, commonly found in simple dense coniferous forests (firs, spruces, and pines), is in the left side of the gradient whereas the western wood pewee, commonly found in more complex configurations in mixed deciduous and coniferous woods and along streams, is on the right side of the graph. *Sources*: https://upload.wikimedia.org/wikipedia/commons/thumb/e/ee/Empidonax_hammondii.jpg/1200px-Empidonax_hammondii.jpg (top left); https://upload.wikimedia.org/wikipedia/commons/f/f1/Western_Wood_Pewee.jpg (top right); Adapted from: McGarigal, K. and McComb, W.C. (1995) Relationship between landscape structure and breeding birds in the Oregon Coast range. *Ecological Monographs*, 65, 235–260 (bottom).

and greater edge contrast. The distribution of 12 species of birds along that axis of increasing complexity of patch shapes and higher edge contrast is shown in Figure 8.6.

It is very easy to see from Figure 8.6 that the bird species are differently distributed along the fragmentation gradient. Some, such as the Hammond's flycatcher (*Empidonax hammondii*), are located on the left part of the graph, indicating their preference for continuous patches of forests with low edge contrast, whereas others, such as the western wood peewee (*Contopus sordidulus*), were more strongly associated with all types of fragmented late-seral forests with more complex shapes and more edge contrast. These results show how landscape indices reflecting the complexity of landscape configuration and contrasting edges can be used to understand habitat use and preferences of different bird species.

8.4 Configuration Assessed by Types of Cell Adjacencies

In the above section we saw how different types of edges could be combined into a single measure for the landscape, the edge contrast index (EDCON) for the whole landscape by giving weights to the different edge types. However, it is often difficult to establish the weights to be given to the different edge types. It is therefore, often better to consider the edge types as being quantitatively different but not necessarily giving weights.

A different approach is then proposed using all unique types of cell adjacencies as being different. In this approach, cells, not patches, are the elements of interest where a given patch may be composed of more than a single cell. These metrics cannot be calculated using vector representations of the landscape. The nature of the adjacency between cells is used to quantify the spatial relationships between and within classes, that is, the pattern.

In this approach, similarly to what was done before where habitat diversity of a landscape (HDL) was applied to measure the composition of a landscape by using the proportions of area occupied by the different habitats, it is possible to consider, in raster representations, the proportions of types of cell adjacencies, including those inside a patch, to measure the configuration of the same landscape.

This approach is very useful to understand the differences between two aspects of landscape pattern or structure: composition and configuration. The example of the two landscapes above (Figure 8.1) will be used again to illustrate the calculations. First, in order to measure the configuration of the first landscape, we start by calculating its adjacency matrix. From left to right we observe 10 adjacencies from cells of class A on the left to cells of class A on the right. We also observe 2 adjacencies from cells of class B on the left to cells of class A on the right, and another 2 cells of class B on the left to cells of class C on the right. Finally, we count 6 adjacencies from class C on the left to class C on the right. This would result in the adjacency matrix shown in Figure 8.7.

We can then make the same calculations from top to bottom and sum those values in a single matrix, as shown in Figure 8.8.

Finally, if we consider that the adjacencies from A to B are equivalent to the transitions from B to A we can sum the number from A to B (1) with the number from B to A (2) and consider only one type of adjacency (A to B) as the sum of the two numbers ($1 + 2 = 3$). Using this procedure (summing the symmetrical values outside the diagonal) we can summarize the adjacencies of the first landscape in a half of a square matrix and we

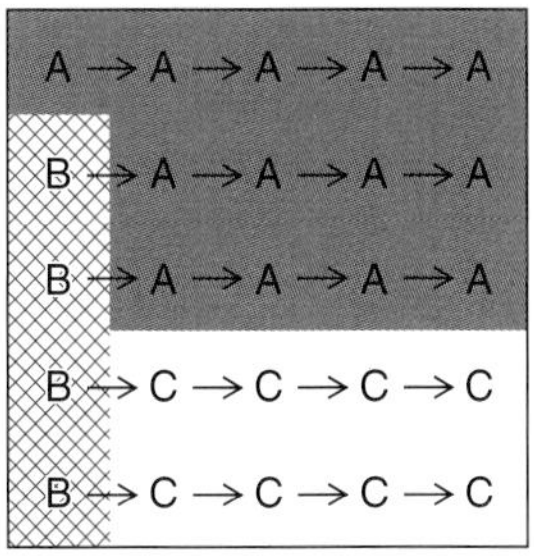

	To Right		
	A	B	C
From Left A	10	0	0
B	2	0	2
C	0	0	6

Figure 8.7 Spatial transitions observed from left to right in the first landscape represented in Figure 8.1, with the corresponding adjacency matrix.

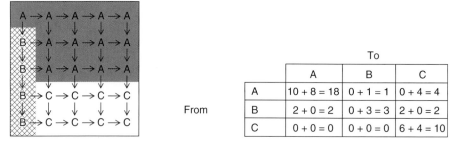

From	To	A	B	C
	A	10 + 8 = 18	0 + 1 = 1	0 + 4 = 4
	B	2 + 0 = 2	0 + 3 = 3	2 + 0 = 2
	C	0 + 0 = 0	0 + 0 = 0	6 + 4 = 10

Figure 8.8 Spatial transitions observed from left to right and from top to bottom in the first landscape represented in Figure 8.1, with the corresponding adjacency matrix.

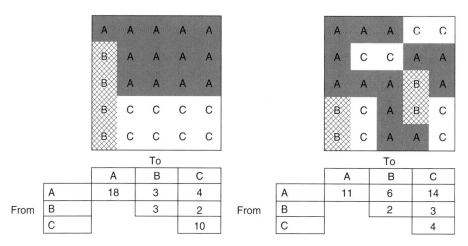

From	To	A	B	C
	A	18	3	4
	B		3	2
	C			10

From	To	A	B	C
	A	11	6	14
	B		2	3
	C			4

Figure 8.9 The two landscapes represented in Figure 8.1 showing the corresponding half of a square adjacency matrix. The diagonal values, indicating like adjacencies, are higher in the first landscape with a simpler configuration.

can do exactly the same calculations for the second landscape. The resulting half of square matrices are shown in Figure 8.9. These adjacency matrices are the basis for most of the calculations of the complexity of the configuration.

For large maps the total frequency in the matrix is approximately the number of cell sides and two times the number of cells. It is always approximate because the border cell sides are not counted. With large numbers of cells in a landscape this difference becomes relatively small. In fact, for a square landscape with $n \times n$ cells, the number of cell sides analyzed is $2 \times n \times (n - 1)$. In the case of our examples, for a small square landscape with $n = 5$, we have 25 cells and 40 adjacencies included in the frequency matrix.

From the adjacency matrices it is very easy to compute the total edge line. In fact, the total edge line (TL) is simply the sum of the nondiagonal elements of the matrix. We can also measure the landscape area (TA) as the number of cells present in the landscape.

As we saw before, the quantity of edge lines in a landscape is a very important measure of complexity of the configuration, but the nature of the edge is also important. To account for the nature of the edge it is possible to calculate the edge diversity of a landscape (EDL) as a measure of the configuration of that landscape by calculating a similar index using the proportions of different edges. The edge diversity of a landscape (EDL) can therefore be computed as

$$EDL = \exp \sum \left[- \left(e_{jk} \right) \ln \left(e_{jk} \right) \right]$$

where e_{jk} is the proportion of the total edge line (TL) that occurs between habitats j and k:

$$e_{jk} = L_{jk}/TL \text{ with } j \neq k$$

The maximum value of this index is the total number of possible edge types, which is calculated as $m*(m-1)/2$, where m is the number of habitats in the landscape. In our case with 3 habitats the maximum number is $3*(3-1)/2 = 3$.

We will go back again to our earlier example to illustrate the calculations. The results are shown in Table 8.2.

It is apparent from the results obtained with the calculations shown in Table 8.2 that the edge diversity of the first landscape is higher than that of the second. In fact, in the first landscape, all edge types are well represented and the value of EDL (2.889) is close to its possible maximum (3). In the second landscape, the diversity of edges is lower because

Table 8.2 Calculations of the edge diversity of the landscape (EDL) for the two landscapes represented in Figure 8.9.

For the first landscape:

Edge line types	Number of cell edges	Proportion (e_{jk})	$-e_{jk} \ln (e_{jk})$
AB	3	0.333	0.366
AC	4	0.444	0.360
BC	2	0.222	0.334
Total edge lines (TLs)	9	Sum	1.061
		Edge diversity of the landscape (EDL)	2.889

For the second landscape:

Edge line types	Number of cell edges	Proportion (e_{jk})	$-e_{jk} \ln (e_{jk})$
AB	6	0.261	0.351
AC	14	0.609	0.302
BC	3	0.130	0.266
Total edge lines (TLs)	23	Sum	0.918
		Edge diversity of the landscape (EDL)	2.505

we have a very high proportion of adjacencies (0.609) between cells of type A and type C and therefore the value of EDL (2.505) is further away from the possible maximum (3). This diversity of edges could be very relevant for many ecological processes as for the diversity of species requiring different types of edges.

We may want an index that is independent of the maximum possible value. We then have the evenness indices that range from 0 when one class is completely dominant to 1 when all classes are equally represented. The edge evenness of the landscape (EEL) could therefore be computed as

$$EEL = EDL/[(m)*(m-1)/2]$$

In our cases we have:

For the first landscape: EEL = 2.889/3 = 0.963
For the second landscape: EEL = 2.505/3 = 0.835

As seen with the edge diversity of the landscape (EDL), the first landscape has a more balanced proportion of edge types and therefore a higher value of EEL, close to its maximum (1).

A similar procedure uses the log values to compute a similar edge evenness index. This is the case of the interspersion and juxtaposition index (IJI) of McGarigal and Marks[6], which is simply another measure of edge evenness and can be computed as

$$\text{Interspersion and juxtaposition index (IJI)} = 100* \ln{(EDL)}/\ln{[(m)*(m-1)/2]}$$

For our examples we have:

For the first landscape: IJI(%) = 100 * ln (2.889)/ln (3) = 96.6
For the second landscape: IJI(%) = 100 * ln (2.505)/ln (3) 83.6

As easily observed, these values are very similar to those of the first index of edge evenness of the landscape (EEL) and provide the same information.

A final and more complete measure of configuration uses all the elements of the adjacency matrix, including the diagonal. This is the adjacency diversity of a landscape (ADL), computed as

$$ADL = \exp \sum \left[-\left(ad_{jk}\right) \ln \left(ad_{jk}\right) \right]$$

where ad_{jk} is the proportion of cell adjacencies that occur between habitats j and k (now using all adjacencies including those within the same habitat $j = k$).

We will use our examples and the adjacency matrices shown in Figure 8.9 again to illustrate the calculations. The results are shown in Table 8.3.

It is apparent from the results of Table 8.3 that the second landscape has a higher value for adjacency diversity since the proportions of the types of adjacencies are a little more balanced than in the first landscape where only the internal adjacencies AA and CC are a very high proportion (0.70) of the total.

Here, again, we can compute an evenness index that is independent of the number of classes in the landscape. If we have m classes, the total number of adjacency types is $m * (m + 1)/2$. In our case, with 3 classes ($m = 3$) the number of possible adjacencies is $3 * (3 + 1)/2 = 6$.

Table 8.3 Calculations of the adjacency diversity of the landscape (ADL) for the two landscapes shown in Figure 8.9.

For the first landscape:

Adjacency types (jk)	Number of cell adjacencies (L_{jk})	Proportion of adjacencies (ad_{jk})	(ad_{jk}) ln (ad_{jk})
AA	18	0.450	0.359
AB	3	0.075	0.194
AC	4	0.100	0.230
BB	3	0.075	0.194
BC	2	0.050	0.150
CC	10	0.250	0.347
Total adjacencies (TL)	40	Sum	1.474
		Adjacency diversity of the landscape (ADL)	4.369

For the second landscape:

Adjacency types (jk)	Number of cell adjacencies (L_{jk})	Proportion of adjacencies (ad_{jk})	(ad_{jk}) ln (ad_{jk})
AA	11	0.275	0.355
AB	6	0.150	0.285
AC	14	0.350	0.367
BB	2	0.050	0.150
BC	3	0.075	0.194
CC	4	0.100	0.230
Total adjacencies (TL)	40	Sum	1.581
		Adjacency diversity of the landscape (ADL)	4.861

Therefore the adjacency evenness of the landscape (AEL) is $ADL/[m * (m + 1)/2]$, or for the first landscape, adjacency evenness (AEL) = 4.369/6 = 0.728 and for the second landscape, adjacency evenness (AEL) = 4.861/6 = 0.810. Again, here we see that configuration is more complex in the second landscape.

Many authors use the complement of adjacency evenness as a measure of contagion[7,8,9] as very often lower values of the adjacency evenness (minimum of 0) may result from landscapes with a few large, contagious patches, whereas higher values (maximum of 1) generally characterize landscapes with many small dispersed patches[10]. This decrease in adjacency evenness with increased contagion resulting in large patches is generally the case, as was observed in the example provided. However, when some adjacencies between cells of different classes are very common (as with adjacencies AC in the second landscape), the value of adjacency evenness also decreases.

Contagion is best measured, as it was done in the previous chapters, as the ratio between like adjacencies (interior of patches) and total adjacencies, that is, the proportion of interior adjacencies (IP). Adjacencies between cells of the same class (like adjacencies or interior adjacencies) are in the diagonal of the adjacency matrix, as adjacencies between cells of different classes (edges) are outside the diagonal of the matrix. The complement of the proportion of interior adjacencies (IP) is the proportion of exterior or edge adjacencies (EP), that is, IP + EP = 1.

In our examples the proportion of interior adjacencies differs in the two landscapes. Whereas in the first landscape we have 31 interior adjacencies and the proportion of interior adjacencies is therefore IP = 31/40 = 0.77, in the second landscape the proportion of interior adjacencies is much smaller: IP = 17/40 = 0.43. This indicates that the first landscape is more associated with large patches than the second landscape.

The proportion of exterior (edge) adjacencies (EP) is the complement of the interior adjacencies (EP = 1- IP) and has exactly the same meaning as edge density, that is, it is a measure of the complexity of the configuration of the landscape.

8.5   Combination of Landscape Pattern Indices

From the previous considerations, it was concluded that various landscape pattern indices derived from cell adjacencies provided indications of different aspects of the configuration of a landscape. It would be important, however, to understand how the various indices are interrelated in order to better interpret their meaning. The fact that many indices are derived from information theory allows for that integration. In fact, Shannon[11] had already stated that an important property of his information measure, H, was that "if a choice be broken down into two successive choices, the original H should be the weighted sum of the individual values of H". If we use the modified Shannon index, the sum is transformed into a product.

In practice, the total adjacency matrix may be broken down into successive choices. The first choice is between like adjacencies (diagonal) and elements above the diagonal. We defined edge proportion (EP) as the proportion of non-diagonal frequencies and interior proportion (IP = 1 − EP) as the proportion of interior adjacencies. An interior/edge diversity index (IEDI) can be computed as a function of the edge and interior proportions as

$$\text{IEDI} = \exp\{-[\text{EP}\ln(\text{EP}) + (\text{IP})\ln(\text{IP})]\}$$

As we have only two types of adjacencies (interior and edge) the value of IEDI is always between 1 (when there is only one type of adjacency, all interior adjacencies, or total edge adjacencies) and 2 (when interior and edge adjacencies are in equal proportions, IP = EP = 0.5).

Then the two subsequent classes (above the diagonal and diagonal elements) can be further subdivided. The diversity of nondiagonal elements is expressed in the index of edge diversity of the landscape (EDL). On the other hand, the diversity of like adjacencies, represented by the diagonal elements, can result in an index of diversity of interiors in the landscape (IDL).

In order to combine the two diversity measures (edge and interior) into a single measure of diversity of whole adjacencies these indices should be weighted by their proportion in the landscape. The final decomposition of the index of adjacency diversity of the landscape (ADL) is

$$ADL = IEDI * EDL^{EP} * IDL^{IP}$$

where IEDI represents the influence of edge/interior proportions, EDL^{EP} represents the influence of edge diversity, and IDL^{IP} represents the influence of the diversity of interior adjacencies. For the two last components landscape composition is obviously important.

When the edge is minimal, IEDI is close to unity, EP is close to zero, and therefore EDL^{EP} is also close to unity, IP is close to unity, and thus the adjacency diversity of the landscape (ADL) is close to the diversity of interior adjacencies (ADL is close to IDL). In this case adjacency diversity would mostly represent composition diversity. On the other hand, when interior adjacencies are minimal, ADL reflects only diversity of the edge types, that is EDL.

For our example we have already calculated for the two landscapes the values for adjacency diversity (ADL) and for edge diversity (EDL). We can now calculate edge and interior proportions (EP and IP), compute the index for interior/edge diversity (IEDI), and the indices of diversity of interior adjacencies for the two landscapes (IDL):

For the first landscape: EP = 9/40 = 0.225 and IP = 31/40 = 0.775 and therefore
   IEDI = exp { − [0.225 ln (0.225) + 0.775 ln (0.775)] } = 1.704
For the second landscape: EP = 23/40 = 0.575 and IP =17/40 = 0.425 and therefore
   IEDI = exp { − [0.575 ln (0.575) + 0.425 ln (0.425)]} = 1.978

The results show that in the second landscape the proportions of edge and interior adjacencies are very close and therefore IEDI approaches its maximum value (2). In the first landscape interior adjacencies are 3 times more abundant than edge adjacencies, showing contagion and a lower value for IEDI.

We can now calculate the diversity of interior adjacencies (IDL) for both landscapes in Figure 8.9, similarly to what was done for the diversity of edges in Table 8.2 and for all adjacencies in Table 8.3. The calculations and the results for the diversity of interior adjacencies (IDL) are shown in Table 8.4.

We can now summarize our indices in Table 8.5.

From the results shown in Table 8.5 we have the following decomposition of the adjacency diversity of the landscapes:

$$ADL = IEDI * EDL^{EP} * IDL^{IP}$$

This is the fundamental equation for the combination of landscape pattern indices.

We then have the following decomposition of adjacency diversity for the first landscape:

$$4.369 = 1.704 * 2.889^{0.225} * 2.476^{0.775}$$

or

$$4.369 = 1.704 * 1.270 * 2.019$$

Table 8.4 Calculations of the diversity of interior adjacencies (IDL) for the two landscapes represented in Figure 8.9.

For the first landscape:

Interior adjacency types (jj)	Number of like adjacencies (L_{jj})	Proportion of like adjacencies (ad_{jj})	(ad_{jj}) ln (ad_{jj})
AA	18	0.581	0.316
BB	3	0.097	0.226
CC	10	0.322	0.365
Total interior adjacencies	31	Sum	0.907
		Diversity of interior adjacencies (IDL)	2.476

For the second landscape:

Interior adjacency types (jj)	Number of like adjacencies (L_{jj})	Proportion of like adjacencies (ad_{jj})	(ad_{jj}) ln (ad_{jj})
AA	11	0.647	0.282
BB	2	0.118	0.252
CC	4	0.235	0.340
Total interior adjacencies	17	Sum	0.874
		Diversity of interior adjacencies (IDL)	2.396

Table 8.5 Summary of indices of configuration calculated using the two representations of landscapes shown in Figure 8.9.

Index	First landscape	Second landscape
Adjacency diversity (ADL)	4.369	4.861
Edge proportion (EP)	0.225	0.575
Interior proportion (IP)	0.775	0.425
Interior/edge diversity (IEDI)	1.704	1.978
Edge diversity (EDL)	2.889	2.505
Interior diversity (IDL)	2.476	2.396

and for the second landscape:

$$4.861 = 1.978 * 2.505^{0.575} * 2.396^{0.425}$$

or

$$4.861 = 1.978 * 1.695 * 1.450$$

From the above comparison it is apparent that the higher value for the complexity of the configuration of the second landscape (ADL 4.861 compared to 4.369 for the first landscape) is related to a more balanced proportion of interior and edge adjacencies (IEDI 1.978 compared to 1.704 for the first landscape), and with a stronger effect of the diversity of edges (1.695 versus 1.270) due to the higher proportion of edges in the landscape (EP 0.575 versus 0.225 in the first landscape) and in spite of having a lower value in the component associated with the diversity of interior adjacencies (1.450 compared to 2.019).

These results are very useful in analyzing the reasons for the differences in complexity of the two landscapes with the same composition used as examples. The differences were very obvious but the quantification of the various indices gives us a means to differentiate landscapes and provide better insights into the ecological importance of those differences. It is also important to note that these components might have very different consequences for different animal or plant species or for ecological processes. Some species and processes will benefit from diversity of interiors, diversity of edges, or total adjacency diversity according to their different natural history needs.

The decomposition of the adjacency diversity of the landscape (ADL) into components provides a coherent means to quantify landscape patterns. It is common, however, to illustrate changes in a landscape pattern by providing a series of indices that show different aspects of the pattern.

8.6 Example of Uses of Pattern and Configuration Metrics to Compare Landscapes

As an example of the common approach to compare configuration of landscapes and to interpret the results we can use the study of the changes in land cover and landscape pattern in northwestern Oklahoma, USA from 1965 to 1995, summarized in Table 8.6.

The authors of the study[12] used several landscape metrics to quantify those changes including the mean patch area (MPA), the mean core area (MCORE), and the mean shape index (MSI), discussed in previous chapters, as well as the total edge line (TL) and the interspersion and juxtaposition index (IJI) discussed in this chapter. Landscapes dominated by woodland vegetation types had greater TL and IJI values but MSI was not affected.

The decreasing patch and core area size and increasing edge and IJI were thought to have major implications for the conservation of grassland vegetation. Many of the grassland biota are sensitive to the minimum core area and the presence of nearby woodland vegetation. The expanding woodlands exacerbated the human-caused influences on grassland fragmentation. The preservation of the remaining areas of native grasslands is a major conservation effort in the Great Plains of North America (Figure 8.10).

Table 8.6 Landscape pattern indices and configuration metrics calculated for 225 different landscapes of northwestern Oklahoma (1965–1995) showing the differences between landscapes with an anthropogenic matrix and those where woody vegetation is dominant.

Landscape index	Landscape matrix type	
	Anthropogenic	Woody vegetation
Mean patch area (ha)	3.76	2.19
Mean core size (ha)	1.81	0.86
Mean shape index	1.71	1.72
Total edge (km)	5.26	7.60
Interspersion/juxtaposition index	55.41	63.38

Adapted from Coppedge, B.R., Engle, D.M., Fuhlendorf, S.D., *et al.* (2001) Landscape cover type and pattern dynamics in fragmented southern Great Plains grasslands, USA. *Landscape Ecology*, 16, 677–690.

Figure 8.10 The Tallgrass Prairie Preserve, in Osage County, Oklahoma, is protected and managed by Nature Conservancy as the largest tract of remaining tallgrass prairie in the world. *Source:* https://upload.wikimedia.org/wikipedia/commons/e/e6/Tallgrass_Prairie_Preserve.jpg.

In a study in the Great Basin of western USA, where the woodland development was more advanced, Bunting and others[13] noted a decrease in IJI as woodland vegetation developed. The IJI decreased as smaller and mid-sized woodland patches coalesced into larger patches of older juniper-dominated woodland. They also noted that these changes threatened the biota of the nearby nonwoodland vegetation.

Different indices reveal different characteristics of landscape configuration. In this chapter we demonstrated how some of the most common indices are calculated and how they are interrelated. We also saw how configuration indices complement composition indices in the assessment of landscape patterns. These indices all refer to landscapes as they are at a specific moment in time. The dynamic nature of landscapes will be considered in the following chapter.

Key Points

- Landscape patterns are important for understanding landscape processes. Landscape patterns can be described by composition and configuration.
- The composition of a landscape is generally measured by different diversity indices. The simplest measure of landscape composition is richness, the number of habitats present in the landscape. Other diversity indices take into account the proportions of the different habitats.
- Landscape diversity is high when landscape classes become equally abundant while landscape diversity is low when one class becomes dominant in the landscape. Landscape evenness, another measure of landscape composition, can be computed by dividing landscape diversity with its richness.
- Landscape configuration describes the arrangement of patches in the landscape. The simplest configuration metrics describe patch size, number of patches, perimeter, and shape. Shape metrics are generally computed as a ratio of perimeter to area.
- Edge contrast metrics differentiate between different types of edge, accounting for the adjacency of different patch types.
- Configuration can be quantified by cell or patch adjacency. These metrics describe the diversity or evenness of edge types. One common index for describing edge type evenness between patches is the interspersion and juxtaposition index (IJI). IJI is high when patches are interspersed, that is, patch types are likely to neighbor all other patch types in the landscape.
- Contagion is a measure of cell adjacency. Contagion is high when cells of the same type are neighboring each other and decreases when cell type interspersion increases.
- Landscape composition and configuration are important for species occupancy and landscape pattern analysis can assist in the understanding of a species habitat needs. Some species prefer large interior patches while others occupy edges or more fragmented landscapes.

Endnotes

1 Forman, R.T.T. and Godron, M. (1986) *Landscape Ecology*, John Wiley & Sons, Inc., New York, NY.
2 Riiters, K.H., O'Neill, R.V., Hunsaker, C.T., *et al.* (1995) A factor analysis of landscape pattern and structure. *Landscape Ecology*, 10, 23–39.
3 Moser, D., Zechmeister, H.G., Plutzar, *et al.* (2002) Landscape patch shape complexity as an effective measure for plant species richness in rural landscapes. *Landscape Ecology*, 17, 657–669.
4 Romme, W.H. (1982) Fire and landscape diversity in subalpine forests of Yellowstone National Park. *Ecological Monographs*, 52, 199–221.
5 McGarigal, K. and McComb, W.C. (1995) Relationship between landscape structure and breeding birds in the Oregon Coast range. *Ecological Monographs*, 65, 235–260.
6 McGarigal, K. and Marks, B.J. (1995) *FRAGSTATS: Spatial Pattern Analysis Program for Quantifying Landscape Structure*, USDA Forest Service General Technical Report PNW-GTR-351, Pacific Northwest Research Station, Portland, OR.

7 O'Neill, R.V., Krummel, J.R., Gardner, R.H., *et al.* (1988) Indices of landscape pattern. *Landscape Ecology*, 1,153–162.

8 Li, H. and Reynolds, J.F. (1993) A new contagion index to quantify spatial patterns of landscapes. *Landscape Ecology*, 8, 155–162.

9 Farina, A. (2006) *Principles and Methods in Landscape Ecology. Towards a Science of the Landscape*, Landscape Series Volume 3, Springer, Netherlands.

10 McGarigal, K., and Marks, B.J. (1995) *FRAGSTATS: Spatial Pattern Analysis Program for Quantifying Landscape Structure*, USDA Forest Service General Technical Report PNW-GTR-351, Pacific Northwest Research Station, Portland, OR.

11 Shannon, C. (1948) A mathematical theory of communication. *Bell System Technical Journal*, 27, 379–423, 623–656.

12 Coppedge, B.R., Engle, D.M., Fuhlendorf, S.D., *et al.* (2001) Landscape cover type and pattern dynamics in fragmented southern Great Plains grasslands, USA. *Landscape Ecology*, 16, 677–690.

13 Bunting, S.C., Strand, E.K., and Kingery, J.L. (2007) Landscape characteristics of sagebrush steppe/juniper woodland mosaics under varying modeled prescribed fire regimes, in *Proceedings of the 23rd Tall Timbers Fire Ecology Conference: Fire in Grassland and Shrubland Ecosystems* (eds R.E. Masters and K.E.M. Galley), Tall Timbers Research Station, Tallahassee, FL, USA.

9

Landscape Dynamics

9.1 The Dynamic Nature of Landscapes: Disturbances and Equilibrium

Landscapes are in a continuous state of flux with their composition and configuration varying constantly temporally. A portion of the temporal landscape change is due to seasonality as changes in organisms' activity affect resource availability[1]. Also changing environmental conditions and resource availability in turn affect other organisms and processes. Additional changes, occurring on a variety of scales, can result from climatic variation, succession, disturbance, or human management of lands[2,3].

Some of the changes involved in landscape dynamics are directional, some reversible, and some appear to occur at random. The landscape composition and configuration at any point results from the sum of two groups of processes: those related to primary and secondary succession (mortality, recruitment, environmental change, soil development) and those related to disturbance (human and natural causes).

Theoretically, a landscape's composition may be at or near equilibrium but the landscape configuration may continue to change through time. This condition is often referred to as dynamic equilibrium or shifting steady-state mosaic[4,5,6] and would occur naturally when the processes related to disturbance are compensated for by those related to recovery or secondary succession when viewed from the appropriate spatial and temporal scale. This condition is obviously extremely scale-dependent. Application of the concept to landscapes is difficult[7] and may apply only when the disturbances are small in a relatively large homogeneous habitat[8,9,10].

The concept of equilibrium landscapes also assumes that the effects of discontinuities or gradients that affect disturbance probabilities and recovery rates are averaged across the landscape, which probably does not occur in real landscapes. Finally, in order to achieve adequate spatial and temporal scales, one risks crossing additional discontinuities[11]. For example, the time scale of succession may overlap that related to climatic change[12,13]. Similar problems occur with the spatial scale. Thus, if equilibrium landscapes occur, they are the exception rather than the rule[11,14]. Equilibrium landscapes may be more a function of humans' desire to find orderliness in the universe than an ecological reality[15].

Applied Landscape Ecology, First Edition. Francisco Castro Rego, Stephen C. Bunting, Eva Kristina Strand and Paulo Godinho-Ferreira.
© 2019 John Wiley & Sons Ltd. Published 2019 by John Wiley & Sons Ltd.

In order to address the problems with spatial and temporal scales, a very influential paper was published by Turner and others in 1993[11] (Figure 9.1). The authors indicate four major factors that determined the disturbance regime of a landscape including:

- Disturbance frequency, or its inverse, the disturbance free interval,
- Rate of recovery following disturbance, or the recovery interval,
- Spatial extent of the disturbance,
- Spatial extent of the landscape.

These factors were then reduced to two key parameters, which represented the effect of space and time on landscape dynamics following disturbance (Figure 9.2). The spatial parameter (on the X axis) was defined as the ratio of the extent of the disturbance to the extent of the landscape. The temporal parameter (on the Y axis) was defined as the ratio of disturbance interval to recovery time. The use of these ratios allowed comparison of landscapes with varying dynamics.

This diagram shows how the spatial and temporal scales are important to understand landscape dynamics. It also shows that the equilibrium or steady state can only be observed when the interval between disturbances is high in comparison with the recovery interval and/or when the system is observed at the appropriate spatial scale where the landscape extent is much higher than the disturbance extent[17].

The question of what is the adequate scale to study landscape dynamics is still open to debate. The concept of the shifting steady-state mosaic suggested by Bormann and Likens[18] to apply to forest landscapes where the proportion of the landscape in each seral

Figure 9.1 Monica Turner in Yellowstone, where pioneer landscape ecology studies were performed. In their concept of equilibrium, disturbance and stability, Turner and others[16] proposed a view of landscape dynamics that included a range of spatial and temporal scales. *Source:* http://www.esa.org/esa/wp-content/uploads/2015/09/MGT_Oct2014_Yellowstone_300dpi.jpg.

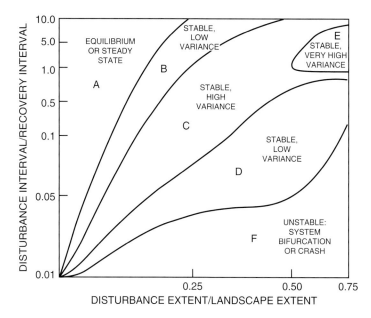

Figure 9.2 Diagram of different types of landscape dynamics as a function of the spatial and temporal parameters. *Source:* Turner M.G., Romme, W.H., Gardner, R.H., *et al.* (1993) A revised concept of equilibrium: disturbance and stability on scaled landscapes. *Landscape Ecology*, 8, 213–227.

stage is relatively constant if averaged over a sufficiently long time or large area has long been studied and seems to be more applicable when disturbances are small and frequent in large areas of homogeneous habitat[19]. However, there is a definitive trade-off between the benefits of increased landscape extent, allowing for a quasi-steady-state landscape where the landscape extent is much larger than the average size of disturbance (>50 times as suggested by Shugart and West[20]), and the disadvantage of increasing landscape extent, which results in the inclusion of greater numbers of different situations in topography, soils, and other factors that can affect the processes of recovery and disturbance.

9.2 The Two-State Landscapes

As seen above, the concept of equilibrium is always set in the context of the disturbance regime and the recovery processes. In the more simple case where only two states are possible, the landscape can be subdivided into two classes:

Class 1: seral stage resulting from disturbance,
Class 2: mature stage developing with ecological succession after disturbance.

In the simplest case, it is possible to model the dynamics of the composition of two-state landscapes by knowing the typical disturbance interval (t_d) and the recovery interval with ecological succession (t_e).

A relative transition matrix (M) between these two states can be built based on the probability that a cell (or a patch) in class 1 changes with ecological succession to class

2 ($e = m_{12}$) and on the probability that a cell (or a patch) in class 2 changes with disturbance to class 1 ($d = m_{21}$).

Assuming that seral stages are not disturbed, we can estimate the probability ($e = m_{12}$) that a cell (or a patch) that is in class 1 (seral stage) will change in one year to class 2 (mature stage) by the probability of having passed enough time to allow for recovery.

We then have

$$e = m_{12} = 1/t_r$$

where t_r is the recovery time (years).

On the other hand, the transition probability (d) that a mature cell (class 2) is disturbed and changes to a seral stage (class 1) is a function of the disturbance interval (t_d). We then have

$$d = 1/t_d$$

We can then develop a relative transition matrix (M) representing this system as shown in Table 9.1.

The results of such a matrix can be further explored in order to know if there is an equilibrium composition of the landscape resulting from those probabilities. Such an equilibrium matrix (EQ) would be a vector with the proportion of the m classes of the equilibrium landscape.

The dynamics of the landscape can be analyzed using the probabilities of changes between classes. The proportion of the seral state (class 1) in the landscape at time t is $p_1(t)$ and would change to time $t + 1$ by adding to the amount that did not recover and remained in the same state $p_1(t) * (1 - e)$, the amount that originated from the disturbed mature state $p_2(t) d$. As $p_1(t) + p_2(t) = 1$ it is obvious that the proportion of class 1 in time $t + 1$ is given by

$$p_1 t + 1 = p_1 t \, 1 - e + [1 - p_1 t] d$$

If there was an equilibrium situation then the proportion of class 1 in the landscape (eq$_1$) would not change in time:

$$p_1(t + 1) = p_1(t) = eq_1$$

We therefore have

$$p_1 = p_1 \, 1 - e + 1 - p_1 d$$

Table 9.1 Matrix corresponding to disturbance and recovery regimes in a two-state landscape.

		To	
		Class 1: Seral	Class 2: Mature
From	Class 1: Seral	Probability to continue in the seral state ($1 - e$)	Probability of recovery from seral to mature ($e = 1/t_r$)
	Class 2: Mature	Probability of disturbance ($d = 1/t_d$)	Probability to continue in the mature state ($1 - d$)

Solving for p_1 we have

$$p_1 = eq_1 = d/(d+e)$$

and solving for p_2 we have

$$p_2 = eq_2 = 1 - p_1 = 1 - [d/(d+e)] = e/(d+e)$$

In matrix terms, the equilibrium landscape (EQ) is a vector with proportions eq_1 and eq_2 that can be calculated using the transition matrix (M) by solving the equation

$$M' \times EQ = EQ$$

where M' is the transpose of M and EQ is the equilibrium vector that has the composition:

$$EQ = \begin{bmatrix} d/(d+e) \\ e/(d+e) \end{bmatrix} = \begin{bmatrix} \text{Proportion of class 1 } (eq_1) \\ \text{Proportion of class 2 } (eq_2) \end{bmatrix}$$

and an associated variance (VARIANCE):

$$VARIANCE = eq_1 \times eq_2 = eq_1 * (1 - eq_1)$$

with a maximum value of VARIANCE $= 1/4 = 0.25$ for $p_1 = p_2 = 0.5$, that is when the equilibrium landscape is composed of equal proportions of seral and mature vegetation classes. This occurs when there are equal probabilities of change in both directions $(d = e)$.

As the probability of disturbance is scale-dependent, the question of the spatial scale considered in the analysis is relevant. The same value of d (and therefore the same equilibrium composition) can result from frequent small disturbances or infrequent large disturbances.

A simple example for gap dynamics in hardwood forests in the Great Smoky Mountains National Park (Figure 9.3) is summarized in Table 9.2 and can illustrate this issue.

Table 9.2 Results allowing for a comparative analysis of the dynamic equilibrium between disturbance and recovery by ecological succession using gap sizes of 75, 250, and 25 m^2 and landscape extents of 100 m^2 and 1 ha (10 000 m^2).

Gap size (m^2)	75	75	250	25
Landscape extent (m^2)	100	10 000	10 000	10 000
Landscape level disturbance interval (years)	100	1	1	1
Annual probability of disturbance (d)	0.0075	0.0075	0.0250	0.0025
Time for recovery (t_r) (years)	91	91	91	91
Annual probability of recovery (e)	0.011	0.011	0.011	0.011
Equilibrium proportion (eq_1)	0.405	0.405	0.694	0.185
Equilibrium proportion (eq_2)	0.595	0.595	0.306	0.815
Variance	0.241	0.241	0.212	0.151

Adapted from Runkle, J.R. (1982) Patterns of disturbance in some old-growth mesic forests of eastern North America. *Ecology*, 63, 1533–1546, and Turner, M.G., Romme, W.H., Gardner, R.H., *et al.* (1993) A revised concept of equilibrium: disturbance and stability on scaled landscapes. *Landscape Ecology*, 8, 213–227.

Figure 9.3 The forested environment of the Great Smoky Mountains National Park, USA. *Source:* https://sasquatchchronicles.com/sighting-by-motorist-in-the-great-smoky-mountains-national-park/.

In the gap dynamics process described a treefall creates a gap of 75 m^2 and occurs in a 100 m^2 plot every 100 years on the average. The recovery time (t_r) for a treefall gap was estimated to be 91 years, the approximate time period required for tree growth to reach the upper canopy. It was also assumed that treefall occurs only in mature forests.

From the comparison of the two first situations we can conclude that the same transition matrix (with the same values of e and d) can originate from different landscape extents with the same disturbance regime. It is also apparent that the situation described in the two first columns indicates a disturbance regime C (lower left to upper right areas in the diagram of Turner and others[21] in Figure 9.2) that is stable (both classes are well represented) and characterized by high or very high variance (the value of 0.241 is very close to its maximum 0.250). The third situation, with a larger gap size (250 m^2), keeping the other variables constant, shows that the equilibrium landscape would be mostly made up by seral stages (eq$_1$ = 0.694) and a lower variance, as those stages tend to be dominant in the landscape and the system might crash (regime F). Finally, when the gap size is small (25 m^2) the landscape would tend to be dominated by mature states corresponding to regimes A or B (equilibrium, steady state, or stable with low variance).

This analysis developed in the context of forest landscapes and is very suitable for forest ecosystems where gap dynamics is prevalent and can be represented by only two states: seral and mature.

9.3 Rotating Landscapes

The concept of a steady-state shifting mosaic conceptually developed for forest ecosystems has nevertheless been traditionally applied in agriculture. In fact, all crop rotation systems have been used for millennia for several reasons, such as breaking pest and

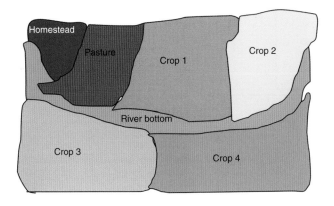

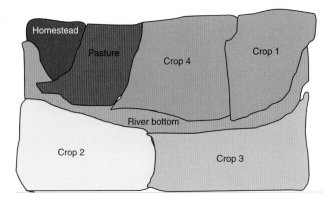

Figure 9.4 Rotational system of traditional English agriculture in dynamic equilibrium, resulting in a very stable landscape composition.

disease cycles, providing straw for bedding or mulching, or adding nutrients or organic matter to the soil.

This rotational system was traditional in European agriculture, where typically each village was surrounded by fields where a different crop was grown every year. Typically one field was allowed to rest with no crops (fallow) to allow the soil to recover. Also, animals could graze in the fallow field so that manure would act as a fertilizer. The dynamics of the resulting landscape can be well illustrated by Figure 9.4.

A rotational system similar to the one of crop succession shown in Figure 9.4 could be simply represented by the 4-year sequence shown in Figure 9.5, where the state A remains for two years, followed by one year of state B, and then one year of state C before returning to state A. We can summarize this system by producing a two-way table with the number of cells that changed from state j to state k. This transition matrix applies for all the one-year periods of the 4-year rotation. The landscape is highly dynamic but its composition is very stable.

From that absolute transition matrix (AT) it is possible to derive a relative proportion matrix (RP) where its elements, rp_{jk}, can be considered as the joint probability of being in class j on the initial landscape and in class k on the final landscape. This joint probability is derived by dividing the at_{jk} elements of the absolute transition matrix (AT) by

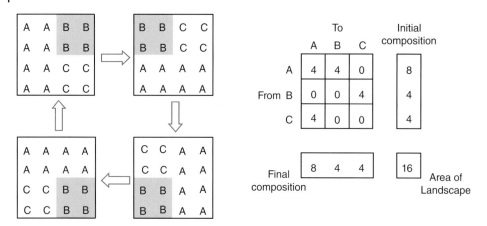

Figure 9.5 Example of a rotational system with a very stable composition (the proportions of the different habitats are constant) but one that is extremely dynamic (a large proportion of cells change class at each time period). The absolute transition matrix (AT) with elements at_{jk} is shown on the right side of the figure.

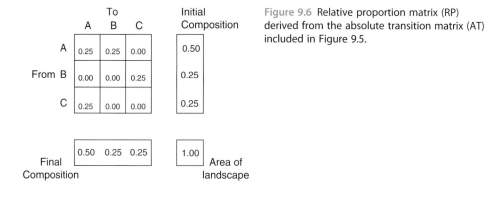

Figure 9.6 Relative proportion matrix (RP) derived from the absolute transition matrix (AT) included in Figure 9.5.

the total of all elements in the table (i.e., the total area of the landscape of 16 cells). With this notation in the relative proportion matrix (RP) the sum of row j represents the proportion of class in time t (the first landscape), $p_j(t)$, and the sum of column represents the proportion of class k, where $p_k(t + 1)$ represents the proportion of class k in time $t + 1$ (the second landscape). The relative proportion matrix (RP) related to our example is shown in Figure 9.6.

Finally, we can define a relative transition matrix (M), where the number of elements for each type of transition is divided by the total number of elements of the row. In this matrix, each element m_{jk} represents the proportion of elements previously in class j in the initial landscape that changed to class k in the second landscape. These elements can also be easily obtained by

$$m_{jk} = rp_{jk}/p_j(t)$$

For our example, the relative transition matrix (M) is shown in Figure 9.7.

This matrix (M) is the most informative of all. From the matrix it can easily be seen that, each year, half of the elements on class A remain in class A and the other half of the

Figure 9.7 The relative transition matrix (M) resulting from the matrices presented in Figures 9.5 and 9.6.

To

		A	B	C
	A	0.5	0.5	0.0
From	B	0.0	0.0	1.0
	C	1.0	0.0	0.0

elements change to class B. It also shows that, each year, all of the elements in B change to C and all the elements in C change to A. The resulting matrix M can be used to generate a model of successional changes. In fact, the landscape composition vector of the landscape in time t can be premultiplied by the transpose of M to generate the landscape composition vector of that landscape in time $t + 1$.

In our example the proportion of the landscape in class A in time $t + 1$ could be calculated as the sum of the m terms referring to the contributions of the m classes:

$$p_A(t + 1) = p_A(t)m_{AA} + p_B(t)m_{BA} + p_C(t)m_{CA}$$

$$p_A(t + 1) = 0.5 \times 0.5 + 0.25 \times 0.0 + 0.25 \times 1.0 = 0.50$$

Similarly, for the proportions of B and C in the new landscape we have

$$p_B(t + 1) = 0.5 \times 0.5 + 0.25 \times 0.0 + 0.25 \times 0.0 = 0.25$$

$$p_C(t + 1) = 0.5 \times 0.0 + 0.25 \times 1.0 + 0.25 \times 0.0 = 0.25$$

This shows that the composition of the landscape at any time $t + 1$ is exactly the same as that at time t. In matrix terms, we have the equilibrium landscape (EQ) as the vector with such proportions that satisfy the equation:

$$M' \times EQ = EQ$$

$$\begin{bmatrix} 0.5 & 0.5 & 0.0 \\ 0.0 & 0.0 & 1.0 \\ 1.0 & 0.0 & 0.0 \end{bmatrix}' \begin{bmatrix} 0.50 \\ 0.25 \\ 0.25 \end{bmatrix} = \begin{bmatrix} 0.50 \\ 0.25 \\ 0.25 \end{bmatrix}$$

In this case EQ is the equilibrium vector with a composition that is half of the landscape in class A ($eq_A = 0.50$), and the remaining half of the landscape divided in equal parts by classes B and C ($eq_B = 0.25$ and $eq_C = 0.25$).

In this situation the agricultural landscape is in a dynamic equilibrium as its composition does not change with time. Obviously, this situation represents the ideal situation of a sustainable landscape that would always maintain its equilibrium. In fact, many factors will intervene that will change this virtual scenario: new pests and diseases, climate change, etc. Nevertheless, it is always important that a reference model is followed as suggested by most agricultural experts.

In fact, typical crop rotation systems proposed by experts today suggest that different crops are grown in the same field in a planned succession, including at least one soil-conservation crop such as perennial hay, which reduces the risk of soil erosion (Figure 9.8). Often legume species or other nutrient fixing species are used in the rotation for some years in an attempt for the soil to recover its previous levels of fertility. This is exactly comparable to the recovery interval presented in the equilibrium model.

Figure 9.8 Conservation crop rotations recommended in the upper Midwest of the USA typically include row crops alternating with small grains and hay or vegetable crops. Conservation crop rotations are most effective when used together with nutrient management as well as upland erosion and runoff control practices such as conservation tillage, contour farming, and grassed waterways. *Source:* https://commons.wikimedia.org/wiki/File:Corn_Fields,_Iowa_Farm_7-13_(15277889101).jpg.

9.4 Indices for the Dynamics and Randomness of Landscape Changes

In practice, landscape change is also normally evaluated by the comparison of two or more maps from differing time periods of this continuously changing entity. These maps may be determined from a variety of methods such as: comparison of remotely sensed images taken at different times, results of field samples and maps, results of landscape model projections or, in many instances, results of a combination of different sources.

The scope of this text does not include methods to develop the maps used in landscape analysis. There are many excellent texts and other works available on the use of various field methods, processing of remotely sensed data, and application of landscape models[22,23]. There are also methods available to assess the accuracy and errors associated with these maps of the landscape. It will be assumed here that the reader has two or more maps of a landscape from different time periods and knowledge of the nature of the metadata that is incorporated into each map.

Landscape dynamics can be assessed by comparing maps representing a landscape at different points in time. Differences between maps can be quantified as an absolute transition matrix (AT) of the specific areas (polygons or pixels) of each class in the first map that is in each class of the second map[24]. If the two maps have the same number of classes (m), an $m \times m$ contingency table can be constructed, as done in our example. We can then derive a relative proportion matrix (RP) by dividing each element of the matrix AT by the total number of elements.

A good measurement of the dynamic nature (or the inertia) of the landscape is a coefficient of inertia (CI) of that landscape, calculated as the sum of the diagonal elements (rp_{jj}) of the relative proportion matrix (RP). We then have

$$CI = \sum rp_{jj}$$

This statistic (CI) can have a value ranging from 0 to 1, with 0 indicating that the landscapes have nothing in common and 1 indicating that the landscapes are identical. It is often interesting to know what would be the value of CI when overlaying two random maps with the proportions found in the landscape at the two moments in time. In this case, we could calculate a random relative proportion matrix (R') with elements $rp'_{jk} = p_j(t)p_k(t+1)$ and the sum of the diagonal elements (CI') would be

$$CI' = \sum rp'_{jj}$$

It is therefore sometimes useful to compute another coefficient of agreement (CA), equivalent to Cohen's kappa-K, which measures the degree of agreement of the two maps above what would be expected only by chance[25,26,27]:

$$CA = (CI - CI')/(1 - CI')$$

Using our example (Figure 9.6) we first calculate the coefficient of inertia between the landscapes at the two moments. The sum of the diagonal elements of the matrix RP (Figure 9.6) is simply the first term and therefore we have

$$CI = 0.25 + 0.00 + 0.00 = 0.25$$

This value of CI (0.25) means that only one-quarter of the area remains unchanged during that period (one year), therefore indicating a very dynamic landscape. It should be emphasized here that a dynamic landscape does not mean necessarily a dynamic landscape composition. In the above example we observe exactly the situation where a highly dynamic landscape has a constant composition.

We can now build the random relative proportion matrix (RP') whose elements are the product of the proportions of the classes in the initial composition of the landscape and the proportions in the final composition. For our example the random relative proportion matrix (RP') is shown in Figure 9.9.

We can now compute CI' as the sum of the diagonal elements of this matrix:

$$CI' = (0.250 + 0.0625 + 0.0625) = 0.375$$

Figure 9.9 The random relative proportion matrix (RP') from the example described.

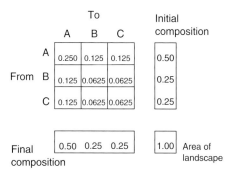

We can then calculate the coefficient of agreement (CA) as

$$CA = (0.250 - 0.375)/(1 - 0.375) = -0.20$$

This value indicates that the landscapes "disagree" in their composition. The coincidences of elements in the same position were much less than those expected from random changes. From the above analyses it can be concluded that, in the period observed (one year), the landscape changed very significantly (only 25% of the elements remained unchanged at CI = 0.25) and that these changes were not random as we would then expect many more elements in the diagonal (CI' = 0.375). The negative value for CA (−0.20) indicates that there are less elements remaining in place than we would expect in a random process.

The results only confirm what we already knew about the process, as we knew what the process was and how the pattern of change originated. However, it is more useful that we detect pattern as a clue to understand the underlying landscape processes.

9.5 Measuring the Complexity of Landscape Change

From the relative proportion matrix (RP), it is also possible to compute several other different metrics. A very useful approach is to view the problem of the comparison of maps (initial and final) through the use of information theory and the concept of mutual information. The average mutual interpretation index provided by Finn (1993)[28] applies such an approach, which can be viewed as the temporal analogue of the indices used to determine the complexity of configurations, where the temporal transition matrix replaces the adjacency matrix, that is, time adjacency replaces spatial adjacency.

The total diversity of the relative proportion matrix (RP) can therefore be used as an index of complexity of the temporal transitions as the diversity of adjacencies was a measure of the complexity of the spatial configuration of the landscape. The only mathematical difference between the two indices is that the matrix for the spatial adjacencies uses a matrix with the diagonal and the half upper-right part of the matrix (as adjacencies between class j and k have the same meaning as transitions between k and j) while the matrix for temporal transitions uses a full square matrix (as the transition from j to k is fundamentally different from the transition from k to j). We can then quantify the complexity of the change by the transition diversity of the landscape (TDL) with the equation

$$TDL = \exp\left\{ \sum_{ij} \left[-rp_{jk} \ln rp_{jk} \right] \right\}$$

This measure quantifies the equivalent number of transitions as the habitat diversity of the landscape (HDL) quantified the equivalent number of classes in the landscape and the adjacency diversity of the landscape (ADL) quantified the equivalent number of types of adjacencies in the landscape.

The minimum value of the TDL = 1, the more simple case, where all the landscape is composed of one class that does not change in time. The maximum value for TDL in a landscape with m classes is m^2, where all transitions are equally represented. This situation corresponds to the maximum complexity of the transitions.

If we want a measure of complexity that is independent of the number of classes we can calculate a measure of evenness, the value of TDL divided by its possible maximum m^2.

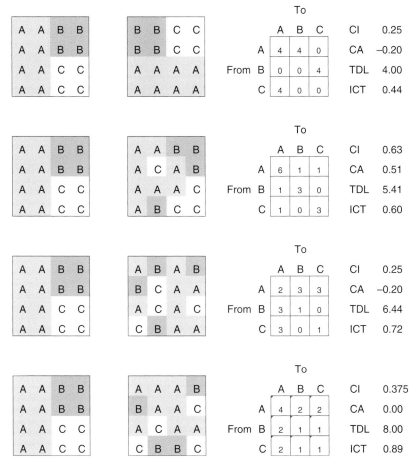

Figure 9.10 Four possible transitions from the same initial landscape resulting in different absolute transition matrices (AT) and different indices of landscape dynamics: the coefficient of inertia (CI), the coefficient of agreement (CA), the transition diversity of the landscape (TDL), and the index of complexity of temporal changes (ICT).

This ratio is a relative measure of the diversity of transitions and can be referred to as an index of the complexity of temporal changes (ICT) with values ranging from 0 to 1:

$$ICT = TDL/m^2$$

In Figure 9.10 we can compare the dynamics and the complexity of four possible transitions from the same initial landscape. In all four transitions shown in Figure 9.10 the initial landscape is the same and the composition of the final landscape is the same. In all cases the landscape composition did not change. However, the process, that is the way classes replace each other in the landscape, is quite different. To illustrate the difference between the transition processes, the two landscapes (initial and final) are presented in pairs, allowing for a better visual perception of the differences. These differences are captured in the corresponding absolute transition matrix (AT) and the resulting indices.

The higher values for CI and CA in the second transition indicate that the corresponding system is the less dynamic one. The increasing value of TDL and ICT from top to bottom indicates that the complexity of the transition changes in that sequence.

The first transition represented in Figure 9.10 summarizes the results done for the rotational system presented before showing a very dynamic landscape where only a small proportion remained unchanged (CI = 0.25), less than that expected for a random situation (CA = −0.20).The indices that refer to the complexity of the transitions (TDL and ICT) both show the lower values of all the transitions presented (TDL = 4.0 and ICT = 0.44). The system is very dynamic and very simple, typical of fast and artificial rotation systems.

The second transition from Figure 9.10 shows a different situation, where many of the elements remained unchanged (CI = 0.75), with positive values for CA = 0.51 well above what would be expected from a random situation, showing more complexity than the first transition (TDL = 5.41 and ICT = 0.60). It is the less dynamic of all the transitions presented.

The third transition in Figure 9.10 is equivalent to the first as the number of diagonal elements is the same, showing a very dynamic situation. It differs from the first situation, however, because the complexity of the transitions is much higher (TDL = 6.44 and ICT = 0.72).

Finally, we have the fourth transition shown in Figure 9.10, where the proportion of elements in the diagonal is exactly equal to what would occur in a random process (CI = 0.375 and CA = 0.00). This transition is by far the more complex (TDL = 8.00 and ICT = 0.89).

These examples illustrate the adequacy of the indices presented to represent the dynamics and the complexity of the transition processes.

9.6 Simulating Changes in Landscape Composition

In the above sections we described the way landscape changes may be quantified in terms of their dynamics and complexity using small artificial examples. Here we will be using real examples of landscape changes observed and analyzed in order to create models that allow the simulation of changes in landscape composition.

An example of this is the Nature Park of Arrábida (Figure 9.11), covering an area of 10.8 thousand hectares, in southern Portugal. This park's creation, very much promoted by a pioneer of ecology in Portugal, Professor Baeta Neves, was aimed at protecting the ecological and cultural values of these beautiful landscapes. In fact, much of its original vegetation cover was threatened by the expansion of quarries and urban sprawl. The creation of the Park aimed at protecting the area from such threats.

The dynamics of the landscapes were possible to study through the use of aerial photographs, which had been taken since 1950, in one of the first studies published on quantifying landscape changes in Portugal using transition models (Figure 9.12)[29].

From the analysis of the changes in the landscape shown in Figure 9.12 with the five classes (Forest, Agriculture, Shrublands, Rocks/beaches, and Quarries/urban) it was possible to derive three relative transition matrices (M) relative to the periods 1950–1980, 1980–1989, and 1989–1994, respectively:

$$M(1950-1980) \begin{bmatrix} 0.43 & 0.01 & 0.51 & 0.01 & 0.04 \\ 0.08 & 0.48 & 0.4 & 0,00 & 0.04 \\ 0.05 & 0.01 & 0.8 & 0.09 & 0.05 \\ 0.02 & 0,00 & 0.47 & 0.49 & 0.02 \\ 0,00 & 0.04 & 0.15 & 0,00 & 0.81 \end{bmatrix}$$

$$M(1980-1989) \begin{bmatrix} 0.74 & 0.03 & 0.21 & 0,00 & 0.02 \\ 0.07 & 0.76 & 0.12 & 0,00 & 0.05 \\ 0.13 & 0.03 & 0.79 & 0.03 & 0.02 \\ 0.03 & 0.01 & 0.52 & 0.39 & 0.05 \\ 0.03 & 0.03 & 0.07 & 0,00 & 0.87 \end{bmatrix}$$

$$M(1989-1994) \begin{bmatrix} 0.46 & 0.03 & 0.49 & 0.02 & 0,00 \\ 0.19 & 0.44 & 0.26 & 0.02 & 0.09 \\ 0.08 & 0.01 & 0.83 & 0.08 & 0,00 \\ 0.01 & 0,00 & 0.46 & 0.53 & 0,00 \\ 0.03 & 0.03 & 0.04 & 0.01 & 0.89 \end{bmatrix}$$

Figure 9.11 Serra da Arrábida. The first attempts to protect the Serra da Arrábida started in the 1940s, culminating in the creation, in 1976, of the Parque Natural da Arrábida. *Source:* Photo by Jcunha123 – Own work, CC BY-SA 3.0, https://commons.wikimedia.org/w/index.php?curid=40252982.

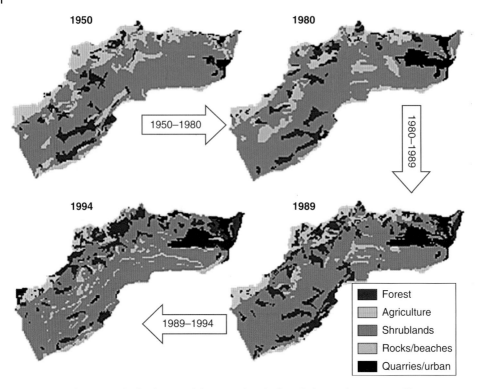

Figure 9.12 Changes in the landscape of the Natural Park of Arrábida, southern Portugal between 1950 and 1994. *Source:* Adapted from Godinho-Ferreira, P., Almeida, M., Fernandes, A., *et al.* (2004) Landscape dynamics in the area of Serra Da Arrábida and the Sado River Estuary, in *Recent Dynamics of the Mediterranean Vegetation and Landscape* (eds S. Mazzoleni, G. Pasquale, M. Mulligan), John Wiley & Sons, Inc., New York, NY.

It is often difficult to visualize easily the differences between different matrices. A better way to look at the transitions is to plot them in such a way that the sum of the proportions from one class to another is displayed as a stacked-bar graph (Figure 9.13). It is clear that, in spite of some differences between the transitions in the three periods, some common trends are very visible.

From this graph it is apparent that, in spite of the different periods and number of years of each period, the three transition periods show many characteristics in common:

• Half of the forests are maintained and half change to shrublands.
• A good proportion of the agricultural fields are maintained but a significant proportion also change to shrublands.
• Shrublands tend to be self-maintained.
• Half of the rocky areas tend to be covered by shrublands.
• Quarries and urban areas tend to perpetuate in the area.

In spite of the obvious similarities, some differences between the transition matrices are also observed. In fact, in the period 1980–1989 forests and agriculture seemed to

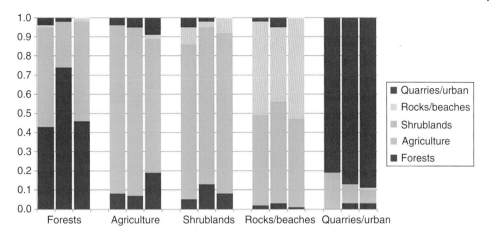

Figure 9.13 Stacked bar graph showing the proportions changed from the class on the X axis to the five classes present in the landscape during the three time periods considered (1950–1980, 1980–1989, and 1989–1994).

change less to shrublands, probably indicating less impacts of wildfires and less agricultural abandonment.

We can now use the same example to simulate the dynamics of the landscape, assuming the constancy of the replacement probabilities over time. The transition from 1950 to 1980 can now be illustrated using the initial composition (vector) in 1950 and the transpose of matrix M:

$$M' 1950 - 1980 \qquad 1950 \qquad 1980$$

0.43	0.08	0.05	0.02	0.00		0.10		0.09
------	------	------	------	------		------		------
0.01	0.48	0.01	0.00	0.04		0.18		0.09
0.51	0.40	0.80	0.47	0.15	×	0.65	=	0.67
0.01	0.00	0.09	0.49	0.00		0.06		0.09
0.04	0.04	0.05	0.02	0.81		0.02		0.06

The proportion of forests in 1980 would be the proportion of forests in 1950 (0.10) times the proportion of forests that remain forests (0.43) plus the proportion of agriculture in 1950 (0.18) that was changed to forests in 1980 (0.08), plus the proportion of shrublands (0.65) that developed into forests (0.05), plus the proportion of rocky areas (0.06) that changed into forests (0.02), plus the proportion of quarries/urban areas (0.02) reconverted to forests (0.00). In summary, the proportion of forests in 1980 (p_{F1980}) would be:

$$p_F 1980 = 0.10x0.43 + 0.18x 0.08 + 0.65x 0.05 + 0.06x 0.02 + 0.02x 0.00 = 0.09$$

The proportions of the other classes are calculated in a similar way.

The simulation of the subsequent landscape compositions over time can be done by successive multiplications. The mathematical process using successive multiplications, referred to as a Markov chain, converges to a stable equilibrium, with a composition where

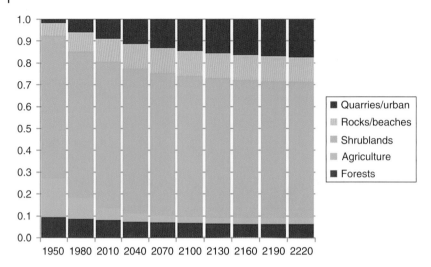

Figure 9.14 The multiplication of the matrix M′ (computed by the transition 1950–1980) by the initial composition vector (1950) results in the landscape composition in 1980. The multiplication of M′ by the composition vector of 1980 gives the prediction for the subsequent time steps. The Y-axis represents the proportions of the different classes in the landscape.

the proportion of each class is independent of the initial composition (Figure 9.14). This stable composition can also be computed in a more elegant way by taking the first eigenvector of M, an operation of matrix algebra that will not be further explained here.

We can make the same calculations for the different periods and have the corresponding equilibrium compositions (Figure 9.15). It is easily concluded that the three

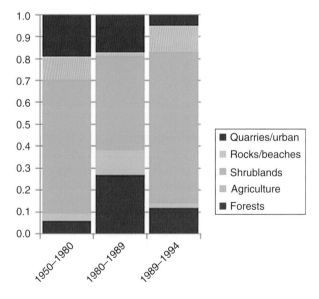

Figure 9.15 Equilibrium composition (EQ) of landscapes resulting from transition matrices (M) observed in different periods (1950–1980, 1980–1989, and 1989–1994).The Y-axis represents the proportions of the different classes in the landscape.

landscapes are similar in that they all predict the dominance of shrublands with variable but smaller contributions of forests and quarries/urban areas.

We can conclude that the general tendency is the same, converging for a landscape mostly composed of shrublands. The period 1980–1989 seemed to correspond to a more balanced situation where forests and agriculture seemed to be able to remain an important component of the landscape, but the following period indicates that there is a tendency for the abandonment of agriculture to shrublands and, possibly because of wildfires, forests are also decreasing and rocks are more exposed. It becomes apparent that the expansion of quarries and urban areas was very much controlled in the period 1989–1994.

This example shows the potential of this simple Markov model to elucidate the nature of the changes in the landscape and to allow for simulations towards a virtual equilibrium landscape. Several other examples of the use of this approach in modeling landscape dynamics in the Mediterranean can be found in the pioneering book edited by Mazzoleni and others[30].

9.7 Conditional Landscape Changes

The examples above illustrated the strict relationship between process (the transition matrix) and pattern (the composition of the equilibrium matrix). We also saw that transition matrices are not necessarily constant through time. In this section we will see that transition probabilities can also be affected by other factors, such as disturbances, and how those factors can be included in this modeling framework.

Wildfires in Portugal provide an example of conditional landscape change (Figure 9.16). Wildfires have been a major disturbance in Portugal and a very significant factor in shaping its landscape. From the 8.9 million ha of the Portuguese mainland during the period 1975–2007 around 1.3 million ha experienced one fire, 0.5 million two fires, and 32 thousand has had 6 or more fires during that 32 year period. In 2017 about half a million hectares burned.

We start by building our relative transition matrix (M) and display it graphically (Figure 9.17), as done before in Figure 9.13. From Figure 9.17 it is possible to observe the average common transitions in the landscape during those 32 years: the urban areas tend to remain in the landscape, some agriculture is abandoned to shrublands, cork, and holm oak open systems "Montado" are also maintained, other broadleaves show various trends, eucalypt and pine forests remain present but maybe at times replaced by shrublands, and shrublands tend to be very stable.

We can now see whether the occurrence of wildfires and the number of times they occurred in the area has had an effect on the transition process. For that we partition the landscape into subareas, each having had the same number of wildfires. Then we develop one relative transition matrix for all areas having the same condition (the number of wildfires occurring in that area). Then we display the results in a similar way to that in Figure 9.17, but showing the results for each fire condition (Figure 9.18). It is easily seen that urban areas do not generally burn and that shrublands tend to dominate when fire frequency increases, except for eucalypts (with considerable regeneration after fire).

We can see in the top left bar that only a very small proportion (0.01) of areas classified as urban had fires in 32 years, and these areas remained essentially unchanged

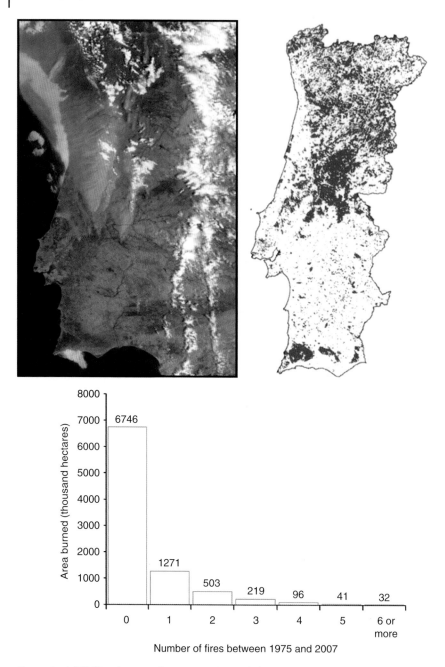

Figure 9.16 Wildfires from satellite in 2003 (upper left), areas burned between 1975 and 2007 (upper right), and area burned in that period in relation to the number of fires that occurred in that area (lower). *Sources*: Google Earth (upper left); J.M.C. Pereira and Portuguese Forest Services (upper right); Marta Rocha, personal communication (lower).

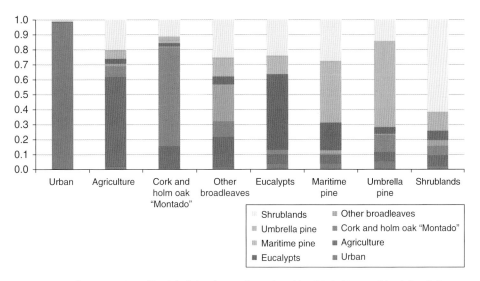

Figure 9.17 The proportions (Y-axis) of the classes that existed in 1975 in Portugal (mainland) that were replaced in 2007 by other classes.

(Figure 9.18). In agricultural areas wildfires were not very frequent (87% of the area did not burn during that period). However, it is observed that the increasing number of wildfires is related to a higher change from agriculture to other classes, in particular to shrublands, probably related to the process of agricultural land abandonment. Cork (*Quercus suber*) and holm oak (*Q. rotundifolia*) open systems "Montados" and forest of other broadleaves, especially deciduous oaks (Figure 9.19), show the same general pattern: the increased frequency of wildfires leads to the increase of shrublands, with other broadleaves experiencing many more wildfires than the "Montados". A completely different situation occurs with eucalypt forests, where the frequency of wildfires seems to have very little influence until fire is very frequent (6 or more fires in the 32 year period). Maritime pine and umbrella pine share the same trends, with very dramatic changes to shrublands. Finally, for shrublands, the higher the frequency the higher the proportion of shrublands remains in the landscape. This class is also the one where wildfire frequency is the highest, as only 46% of the shrubland area did not have any wildfire during that period.

It is clear that the value for the transitions for the whole class is the weighted average of the transitions for each fire frequency and the weights are the proportions of the areas in each fire frequency. It is no surprise that, in our example, the values for the whole class are similar to the values of the class with no wildfire, especially for those classes where fires are not frequent.

The elements m_{jk} of the relative transition matrix (M) shown at the beginning are therefore the weighted average of the various elements (m_{jkv}) of the conditional relative transition matrices associated with the various possible conditions (w_v). We then have

$$m_{jk} = \sum \left(w_v * m_{jkv} \right)$$

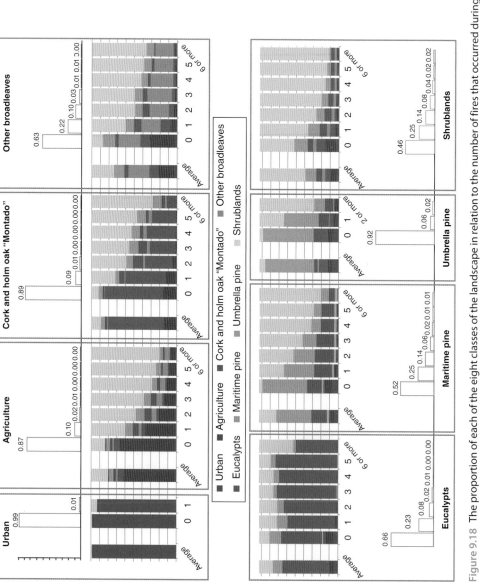

Figure 9.18 The proportion of each of the eight classes of the landscape in relation to the number of fires that occurred during the period 1975–2007 in mainland Portugal (bar graphs at the top and the bottom) and the consequences for the subsequent landscape composition in 2007 (stacked bar graphs in the middle). Each box above indicates the initial class in 1975.

Figure 9.19 The eight landscape cover classes commonly recognized in Portugal. In the top row, left to right: urban (although dispersed houses also burn in wildfires) and agriculture; second row, left to right: cork oak (*Quercus suber*) and Lusitanian oak (*Quercus faginea*). In the third row from left to right: eucalypts (*Eucalyptus globulus*) and maritime pine (*Pinus pinaster*); and in the bottom row, left to right: umbrella pine (*Pinus pinea*) and shrublands (where fire can occur as a "bad master" – wildfires – or as a "good servant" - prescribed fires). *Sources*: Photographs courtesy of João Pinho, Paulo Fernandes, and Conceição Colaço.

Table 9.3 Calculations showing that the proportion of the whole area of maritime pine that changed to shrublands ($m_{jk} = 0.267 = 352\ 270$ ha/1 319 544 ha) can be obtained by the weighted average of the proportions (conditional probabilities) of the area of maritime pine that changed to shrublands under a certain condition (m_{jkv}) where the weights are the proportions of the area of maritime pine in that condition (w_v).

Condition: number of wildfires (v) in the period 1975–2007	Area of maritime pine in condition v (ha)	Weight: proportion of the area of maritime pine in condition v (w_v)	Area changed to shrublands in condition v (ha)	Proportion of the area changing to shrublands under condition v (m_{jkv})	Product $w_v * m_{jkv}$
0	687 667	0.521	71 779	0.104	0.054
1	328 344	0.249	119 586	0.364	0.091
2	180 292	0.137	85 133	0.472	0.065
3	77 126	0.058	43 711	0.567	0.033
4	28 762	0.022	19 264	0.670	0.015
5	10 334	0.008	7 283	0.705	0.006
6 or more	7 017	0.005	5 514	0.786	0.004
	Total maritime pine area (ha) 1 319 544		Total area changing from maritime pine to shrublands (ha) 352 270		Weighted average $m_{jk} = \sum (w_v)(m_{jkv})$ 0.267

We can now compute an example of the transitions between maritime pine (j) and shrublands (k), using all of the different fire frequency conditions (v). The results are shown in Table 9.3.

We see that the element m_{jk} of the average relative transition matrix may be computed directly by dividing the total area that changed from Maritime pine to shrublands by the total area of Maritime pine or, alternatively, as the weighted average of the various m_{jkv} values.

This shows that the relative transition matrix (M) may be decomposed into two components, one (m_{jkv}) related to the transition from class j to class k when the condition v is present (a component that is constant as it depends only on the system) and the other (w_v) related to the probability that the condition v occurs. We will further explore this matter of conditional probability in the next chapter.

Key Points

- Landscapes are continuously changing temporally in composition and configuration. These changes are a result of many interacting factors including: seasonality, organism activity, resource availability, succession, disturbance, and human management activity. The spatial and temporal scale of this dynamic landscape varies greatly.

- Theoretically the landscape composition may be relatively stable at particular spatial and temporal scales. This has been referred to as "dynamic equilibrium" or a "shifting steady-state mosaic".
- The concept of a steady-state shifting mosaic has been traditionally applied in agriculture, where production of several crops is rotated among several different parcels. It has also been applied in forested systems where forest growth, harvest and reestablishment are rotated among many different units within a forested landscape.
- Indices have been developed to quantify the nature and complexity of landscape change that are capable of differentiating between areas with varying disturbance potential or human-caused differences. These indices are useful in predicting landscape composition in the future.

Endnotes

1 Merriam, G., Henein, K., and Stuart-Smith, K. (1991) Landscape dynamic models, in *Quantitative Methods in Landscape Ecology* (eds M.G. Turner and R.H. Gardner, Springer-Verlag, New York, NY.

2 Baker, W.L. (1992) Effects of settlement and fire suppression on landscape structure. *Ecology*, 73, 1879–1887.

3 Vos, W. and Storelder, A. (1992) *Vanishing Tuscan Landscapes. Landscape Ecology of a Submediterranean–Montane Area (Solano Basin, Tuscany, Italy)*, Purdoc Scientific Publishers, Wageningen, The Netherlands.

4 Bormann, F.H. and Likens, G.E. (1979) *Pattern and Process in a Forested Ecosystem*, Springer-Verlag, New York, NY.

5 Sprugal, D.G. (1976) Dynamic structure of wave-regenerated *Abies balsamea* forests in the northeastern United States. *Journal of Ecology*, 64, 889–911.

6 Zackrisson, O. (1977) Influence of forest fires on the North Swedish boreal forest. *Oikos*, 29, 22–32.

7 Turner, M.G., Romme, W.H., Gardner, R.H., *et al.* (1993) A revised concept of equilibrium: disturbance and stability on scaled landscapes. *Landscape Ecology*, 8, 213–227.

8 Zedler, P.H. and Goff, F.G. (1973) Size-association analysis of forest successional trends in Wisconsin. *Ecological Monographs*, 43, 79–94.

9 Pickett, S.T.A. and White, P.S. (1985) Patch dynamics: a synthesis, in *The Ecology of Natural Disturbance and Patch Dynamics* (eds S.T.A. Pickett and P.S. White), Academic Press, New York, NY.

10 DeAngelis, D.L. and Waterhouse, J.C. (1987) Equilibrium and nonequilibrium concepts in ecological models. *Ecological Monographs*, 57, 1–21.

11 Turner, M.G., Romme, W.H., Gardner, R.H., *et al.* (1993) A revised concept of equilibrium: disturbance and stability on scaled landscapes. *Landscape Ecology*, 8, 213–227.

12 Webb III., T. (1981) The past 11,000 years of vegetational change in eastern North America. *BioScience*, 31, 501–506.

13 Delcourt, H.R., Delcourt, P.A., and Webb, T. (1983) Dynamic plant ecology: the spectrum of vegetational change in space and time. *Quaternary Science Review*, 1, 153–175.

14 Pickett, S.T.A. and White, P.S. (1985) Patch dynamics: a synthesis, in *The Ecology of Natural Disturbance and Patch Dynamics* (eds S.T.A. Pickett and P.S. White), Academic Press, New York, NY.

15 Botkin, D.B. (1990) *Discordant Harmonies: A New Ecology for the Twenty-First Century*, Oxford University Press, Oxford.

16 Turner, M.G., Romme, W.H., Gardner, R.H., *et al.* (1993) A revised concept of equilibrium: disturbance and stability on scaled landscapes. *Landscape Ecology*, 8, 213–227.

17 Turner, M.G., Romme, W.H., Gardner, R.H., *et al.* (1993) A revised concept of equilibrium: disturbance and stability on scaled landscapes. *Landscape Ecology*, 8, 213–227.

18 Bormann, F.H. and Likens, G.E. (1979) *Pattern and Process in a Forested Ecosystem*, Springer-Verlag, New York.

19 Pickett, S.T.A. and White, P.S. (1985) Patch dynamics: a synthesis, in *The Ecology of Natural Disturbance and Patch Dynamics* (eds S.T.A. Pickett and P.S. White), Academic Press, New York, NY.

20 Shugart, H.H. and West, D.C. (1982) Forest succession models. *BioScience*, 30, 308–313.

21 Turner M.G., Romme, W.H., Gardner, R.H., *et al.* (1993) A revised concept of equilibrium: disturbance and stability on scaled landscapes. *Landscape Ecology*, 8, 213–227.

22 Sklar, F.H. and Costanza, R. (1991) The development of dynamic spatial models for landscape ecology: A review and prognosis, in *Quantitative Methods in Landscape Ecology* (eds M.G. Turner and R.H. Gardner), Springer-Verlag, New York, NY.

23 Turner, M.G. and Dale, V.H. (1991) Modeling landscape disturbance, in *Quantitative Methods in Landscape Ecology* (eds M.G. Turner and R.H. Gardner), Springer-Verlag, New York, NY.

24 Story, M. and Congalton, R.G. (1986) Accuracy assessment: A user's perspective. *Photogrammetric Engineering and Remote Sensing*, 52, 397–399.

25 Cohen, J. (1960) A coefficient of agreement of nominal scales. *Educational and Psychological Measurement*, 20, 37–46.

26 Congalton, R.G., Oderwald, R.G., and Mead, R.A. (1983) Assessing Landsat classification accuracy using discrete multivariate analysis statisitical techniques. *Photogrammetrc Engineering and Remote Sensing*, 49, 1671–1678.

27 Finn, J.T. (1993) Use of the average mutual information index in evaluating classification error and consistency. *International Journal of Geographical Information Systems*, 7, 349–366.

28 Finn, J.T. (1993) Use of the average mutual information index in evaluating classification error and consistency. *International Journal of Geographical Information Systems*, 7, 349–366.

29 Godinho-Ferreira, P., Almeida, M., Fernandes, A., *et al.* (2004) Landscape dynamics in the area of Serra Da Arrábida and the Sado River Estuary, in *Recent Dynamics of the Mediterranean Vegetation and Landscape* (eds S. Mazzoleni, G. diPasquale, M. Mulligan, *et al.*), John Wiley & Sons, Inc., New York, NY.

30 Mazzoleni S., di Pasquale, G., Mulligan, M., *et al.* (eds) (2004) *Recent Dynamics of the Mediterranean Vegetation and Landscape*, John Wiley & Sons, Inc., New York, NY.

10

From Landscape Ecology to Landscape Management

In the previous chapters we have discussed many ecological principles as they apply to landscape ecology. Among those, the ecological consequences of landscape pattern to landscape processes have been a focus of many chapters. We have also identified numerous metrics to assist in the detection and interpretation of pattern. In this chapter we will give examples of the application of landscape ecology principles and tools such as models to the management of landscapes – in other words, taking it from the realm of scientific theory and concepts to applied ecology of landscapes. These landscapes may vary in size, management objectives and temporal scale, and degree of human control and intervention ("naturalness"). The landscapes used in examples vary from those that are dominated by human activity to those that function under more "natural" conditions. The principles still apply although their application will vary.

10.1   Natural Processes and Landscape Management

Understanding landscape patterns, the dynamic nature of the landscapes and the ecological and social processes that drive landscape changes are all very important aspects of Landscape Ecology.

We have discussed how landscape pattern and process is always associated with landscape dynamics. These concepts can also be applied to landscapes where changes are mostly driven by natural processes or are primarily determined by humans in intensively managed landscapes.

In many cases management attempts to use natural processes as much as possible. This is, for example, the concept of Close-to-Nature Forestry advocated by PROSILVA, an association of foresters who support strategies that use and adapt ecological processes in forest management as the means for rational, sustainable, and profitable management. This approach has been applied mostly in temperate forests of Western and Central Europe (Figure 10.1) and originates from the knowledge that in the past most of Europe was covered by highly complex forest ecosystems with many plant and animal species and that during the recent millennia man has seriously reduced diversity or modified those forests. Close-to-Nature Forestry rejects the treatment of forests as agricultural crops and tries to ensure sustainable and profitable production by imitating the structure and the dynamics of natural forests[1].

Applied Landscape Ecology, First Edition. Francisco Castro Rego, Stephen C. Bunting, Eva Kristina Strand and Paulo Godinho-Ferreira.
© 2019 John Wiley & Sons Ltd. Published 2019 by John Wiley & Sons Ltd.

Figure 10.1 Typical deciduous forests in Western and Central Europe where the concept of Close-to-Nature Forestry first developed on mixed forests using natural regeneration and fine-scale forestry. *Source:* https://upload.wikimedia.org/wikipedia/commons/d/df/Beech_forest_vtacnik.jpg.

However, in many other regions of the world where the impact of man has been more influential and associated with disturbances, the idea of natural forests is much less developed. Possibly one of the best examples in the World of using natural processes to manage landscapes is in Western Australia, where fire has been extensively used by the indigenous people (Figure 10.2).

After the arrival of Europeans in Australia, the historical use of fire by indigenous people was largely abandoned and policies of fire exclusion were established. However, in areas with a Mediterranean climate of hot and dry summers, wildfires (bushfires) were an increasing threat. In 1961, after major wildfires, the Royal Commission recommended the Forests Department to carry out more research into fire control. Foresters in Western Australia, such as McArthur and others, then started developing guidelines for an effective use of prescribed burning[2,3]. The technique of aerial ignition was developed to apply low-intensity prescribed fires for fuel reduction over large forest areas (Figure 10.3).

Under this system of landscape management, using natural processes (fire) similar to what indigenous people had been doing for more than fifty thousand years, landscapes became less homogeneous, composed of patches with various ages after fire, with different fuel loads, therefore making it more difficult for a wildfire to find a percolating patch and making fire-fighting easier (Figure 10.4).

After fuel reduction by prescribed fire, the wildfire control was much more effective, contrary to what happened in southeastern Australia and Tasmania, where the wildfire threat continued. The differences in the success of wildfire control were rightly attributed to fuel reduction, with burning undertaken much more extensively in Western Australia than in Tasmania[4].

Two studies[5] indicate this change from a fire regime dominated by cyclic large wildfires to a system where prescribed fire is applied annually. A 53-year study[6] (1937–1990) in the Perup area east of Manjimup shows a sharp decline of wildfires after the introduction of prescribed burning by the Forests Department in the 1950s. Another 50-year study (1937–1987) in the Collie District of the Department of Conservation and Land

Figure 10.2 Indigenous Australians using fire to hunt kangaroos, circa 1817. *Source:* Painting by Joseph Lycett, National Library of Australia, http://www.nla.gov.au, http://tlf.dlr.det.nsw.edu.au/learningobjects/Content/R4029/object/resource/an2962715s20_nla.jpg.

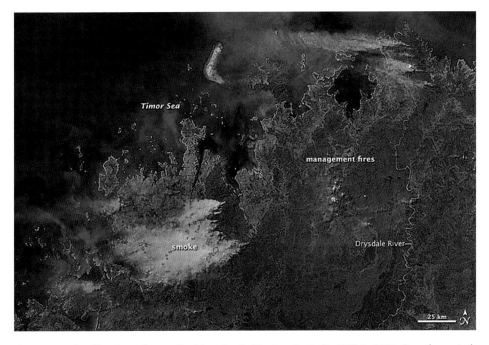

Figure 10.3 Satellite view of prescribed burning in Western Australia (WA) in 2012. Over the period 1963–1990, the whole of the State forest area of the southwest WA progressively came under this science-based system of fire management using lower intensity prescribed fires. *Source:* NASA Earth Observatory, https://en.wikipedia.org/wiki/2011%E2%80%9312_Australian_bushfire_season.

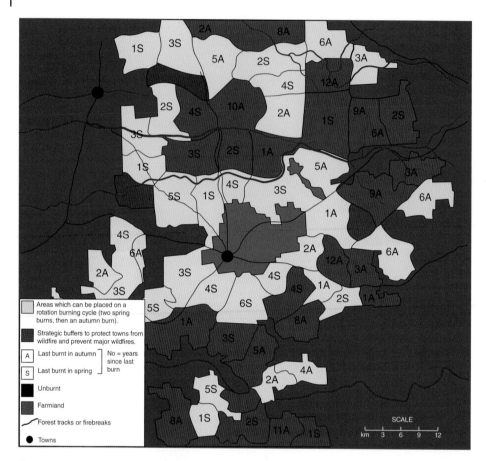

Figure 10.4 The Annual Prescribed Burn Plan 2016–2017 of a forest area near Perth, WA. The landscape mosaic that is created by prescribed fires in different years is described by the various colors on the map. *Source:* Adapted from Burrows, N. (2000) Seasoned with fire, in *Landscape. Fire the Force of ILfe*, Special Fire Edition, Department of Conservation and Land Management, Perth, Western Australia.

Management dominated by jarrah forests (*Eucalyptus marginata*) indicates that prescribed burning practices increased significantly in the 1960s with aerial burning and remained high (9–15% of the area burned annually by prescribed fire), also resulting in a rapid decline in the area burned by wildfires[7] (Figure 10.5).

These examples show how landscapes can be managed to achieve specific objectives, in this case the reduction of wildfire burned area, by simulating natural processes and historical disturbance regimes such as fires. In Western Australia, where the Department of Environment and Conservation manages around 2.5 million hectares and targets at burning annually 200 thousand hectares of forests (8% of the total landscape), summer high-intensity wildfires were being gradually replaced since the 1960s by frequent lower-intensity prescribed fires in spring and autumn. Ecological studies are the basis for many of the operations. Historical and science-based management of these ecosystems follow similar rules based on natural processes.

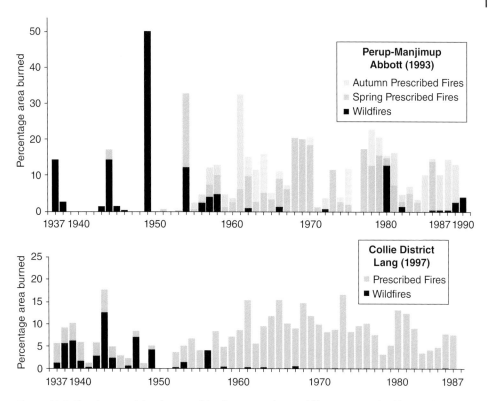

Figure 10.5 The change of dominance of the fire regime from wildfires to prescribed fires in the records of area burned from 1937 to 1990 in two forest areas of Western Australia (Perup-Manjimup in the top graph and the Collie District in the bottom). The figure is adapted from the works of Abbot[8] and Lang[9].

10.2 Transition Matrices as the Mathematical Framework

We saw in the previous chapter how landscape dynamics could be modeled using transition matrices. Here we will see how management can be incorporated in the same framework.

We can look at a landscape as a mosaic composed of different states (classes or habitats) that replace each other in time. If this replacement pattern is such that one state is only replaced by another state then we have a simple structure, as we saw in the previous chapter in the example of rotations.

A simple transition model of this kind was first proposed for forest management by Usher[10,11] and for forest succession by Horn[12]. For both cases the transition model approach could give a simplified version of a relative transition matrix (M) of the form:

$$
M = \begin{bmatrix}
1-e & e & 0 & 0 \\
d & 1-d-e & e & 0 \\
d & 0 & 1-d-e & e \\
d & 0 & 0 & 1-d
\end{bmatrix}
$$

where *e* can be seen as the probability that an area that is in a certain class will be replaced by ecological succession by the next class (probability considered equal for all classes) and *d* as the probability that an area in a certain class is disturbed and is replaced by a pioneer initial state (also considered constant in this example).

This simple formulation has very interesting properties. The parameter *e* can be seen as a measure of the dynamic stability of the ecological succession system and largely controlled by natural factors (as tree growth or biomass accumulation) whereas the parameter *d* can be seen as the probability of disturbance, which could be of natural origin (tree mortality due to age, natural stand replacing fires) or caused by humans (by changing the probability of disturbance and by harvesting).

From this simple formulation it can be seen that, without disturbance (*d*) and without ecological succession (*e*), the landscape would have the maximum inertia with the diagonal equal to unity. We also saw that we could simulate the dynamics of landscape composition by premultiplying the vector of the initial composition V(*t*) by the transpose of M to have the landscape composition vector at time *t* + 1, that is V(*t* + 1):

$$M' V(t) = V(t + 1)$$

Predictions using transition matrices can then be made with the assumptions that the matrix M, that is all the transition probabilities, remains constant over the projection period and that the future state depends only on the current state and not on past history. These are the assumptions of the first-order Markov chains (Figure 10.6).

Figure 10.6 Andrey Markov (1856–1922), a Russian mathematician known for the work on stochastic processes, later known as Markov processes generating Markov chains. *Source:* https://commons.wikimedia.org/wiki/File%3AAndrei_Markov.jpg.

Under those assumptions a Markov process of landscape changes would always converge to a certain landscape composition (EQ) that is in equilibrium with that process. In mathematical terms:

$$M' \text{ EQ} = \text{EQ}$$

In reality, this equilibrium state might never be achieved because of changes in the process (matrix M), but it provides a very good reference for the long-term perspective.

Using our simplified matrix we can predict what would be the equilibrium matrices for different values of the disturbance parameter (d) and the dynamic stability parameter (e). It becomes clear that, for any given value of e (the dynamic parameter), low disturbance probabilities (low values of d) would make all the landscape converge towards the last state and the equilibrium landscape would be composed of only that state. On the contrary, with higher values of d (higher disturbance probabilities) the landscape would tend to converge towards the initial state and the equilibrium landscape would be again very much dominated by only one state.

It is possible then to compute different diversity values (habitat diversity of the landscape, HDL) for different values of the two parameters d and e. As demonstrated by Horn[13], the value of the diversity of the equilibrium composition depends on the number of classes, but it has always a maximum value at an intermediate value of disturbance as measured by a disturbance/stability index (DSI):

$$\text{DSI} = d/(d + e)$$

In the same study, Horn also indicates a simple way to calculate the diversity of a landscape composition based on the disturbance/stability index (DSI), using the modified Simpson index $H_2 = 1/\sum p_j^2$:

$$H_2 = (2 - \text{DSI})/\left[\text{DSI} + 2\,(1 - \text{DSI})^{2m-1}\right]$$

From this equation it can be seen that, for a landscape with only two classes ($m = 2$), the equation reduces to $H_2 = (2 - \text{DSI})/(\text{DSI} + 2)$ with a maximum value of 2 when DSI = 0.5, that is, when the disturbance parameter (d) is equal to the dynamic parameter (e). This is exactly equivalent to what was shown in the previous chapter on disturbance and equilibrium in the two-state landscapes.

The above equation also shows that, for the same value of DSI, the diversity of the landscape increases with its number of classes, or its richness ($H_0 = m$). The equation also shows that with a higher number of classes the maximum diversity is attained with lower values of DSI, showing the importance of disturbance to maintain diversity in landscapes with few classes.

Using our example with four classes we can now calculate two different measures for the diversity of the composition of the equilibrium landscapes, based on either the Shannon or Simpson formulations. The values are plotted in Figure 10.7.

The results of this simple formulation of the problem are in agreement with the intermediate disturbance hypothesis, which indicates that maximum diversity is generally associated with intermediate levels of disturbance[14,15].

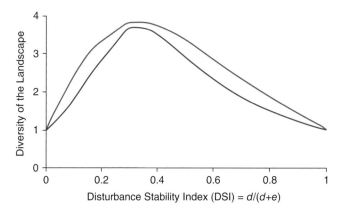

Figure 10.7 The habitat diversity of the landscape (H_1 or HDL) based on the Shannon index (upper line in blue) and the modified Simpson index (H_2 lower line in red) for a landscape with four classes, plotted against the disturbance/stability index proposed by Horn[13]. Higher values of diversity are at intermediate levels of disturbance.

10.3 Management of Landscape Composition and the Transition Matrix Model

The transition matrix model was first proposed for animal populations by Leslie in 1945[16] where the composition of the populations was based on age classes.

A similar concept was proposed in 1966 by Usher, to respond to the needs of the manager of the forest who wanted to know which forest structure would give him "the greatest production, but yet conserve his forest"[17]. In this case the classes were established in terms of size rather than age and, in the case that the units are single trees, size was generally expressed as diameter at breast height (dbh).

In general for the management of forest landscapes or broad-scale analysis, it is more convenient to use areas of forest stands (patches) as units. In this case it is more common that the size of the forest is expressed by a measurement that is more readily used by the manager interested in producing timber: the volume per unit area (typically m^3/ha). Also, it is generally preferred to assume that the transition parameters are different from class to class[18].

In view of the simplicity of the approach based on age classes and that classes reflect the volume available for harvesting, forest simulators of forest dynamics using matrix models are considered more useful to establish classes based on a combination of age classes and volume classes. This is the approach developed by Sallnäs for Swedish forests[19] and the European Forest Information Scenario model (EFISCEN) developed to project future scenarios for the forests in all of Europe[20,21] (Figure 10.8). Figure 10.8 allows for the understanding that some of the processes will occur largely independent of management (e.g., growth, aging, natural mortality) while others are influenced by management regimes (thinning, harvest, artificial regeneration).

Based on transition matrix models with a five-year time step, EFISCEN provides predictions on the forest resource structure, stem wood volume, wood harvest, and, by using conversion factors, biomass and carbon stocks.

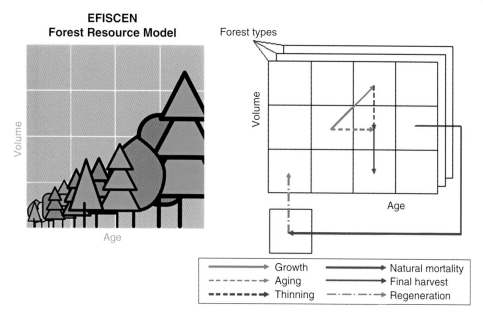

Figure 10.8 The EFISCEN representation of the classes as combinations of age and volume showing the processes that may affect the transitions from one class to the other. Different transition probabilities are defined for each forest type. *Source:* http://www.efi.int/portal/virtual_library/databases/efiscen/.

Recent applications of transition matrix models were proposed and applied using plot data of the Austrian National Forest Inventory[22] and also using an area-based matrix model but with classes defined by combinations of tree density (stem number per hectare) and volume (growing stock in m^3/ha). The development of a European Forest Dynamics Model (EFDM) used classes as combinations of age and volume or combinations of stem density and volume with examples from five countries (Austria, France, Sweden, Finland, and Portugal), concluding the feasibility of this modeling approach, especially for tackling issues traditional models have difficulties with, such as uneven-aged forestry or management under risk[23].

Similar approaches have been used with different objectives, such as optimizing economic returns, tree diversity, or carbon sequestration, and have been applied in many different regions of the world[24,25,26,27]. A good overview and outlook of matrix models in forest dynamics was provided in 2013 by Liang and Picard[28].

The possible format of a corresponding relative transition matrix (M) could be

$$M = \begin{bmatrix} 1-d_1-e_1 & d_1+e_1 & 0 & 0 \\ d_2 & 1-d_2-e_2 & e_2 & 0 \\ d_3 & 0 & 1-d_3-e_3 & e_3 \\ d_4 & 0 & 0 & 1-d_4 \end{bmatrix}$$

where the probability of a tree transition to the first stage (d_i) can include natural regeneration resulting from disturbance and/or artificial regeneration; it is dependent on the class (i) and the probability to change to the next size class (e_i) is dependent on growth and also on harvesting.

10.4   The Use of Transition Matrices to Incorporate Changes in Disturbance Regimes and/or Management Activities

We will use a simple example to illustrate the use of transition matrices to simulate changes in disturbance regimes and to predict future landscape composition resulting from management activities.

As discussed in Chapter 6, in the Owyhee Mountains of southwestern Idaho, USA, a historical quasi-stable composition of the steady-state shifting mosaic has been altered by twentieth century direct and indirect fire suppression actions by land managers. This has resulted in an increase in the area dominated by western juniper woodland and a decrease in the area dominated by sagebrush steppe. These landscape composition changes have resulted in a more homogeneous landscape with a loss of the early and mid seral vegetation stages and a loss of the sagebrush steppe habitat that is critical for the conservation of many species. The photos in Figure 10.9 illustrate the composition changes that have been and are occurring on a broad scale.

In the simplest case (baseline) the transition model would be represented by changes of classes according to a replacement sequence along a successional development and a reverse process caused by wildfire setting back succession to the first successional stage. This process can be represented diagrammatically in Figure 10.10.

Figure 10.9 Photo sequence showing the development of juniper woodland within a sagebrush steppe in North America. Initial juniper woodland (upper left foregrund), advanced juniper woodland (upper right), and old juniper woodland (lower). Photos by Stephen Bunting.

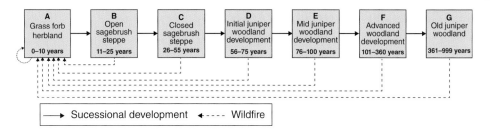

Figure 10.10 Diagram showing the successional development from a grass forb herbland state (A) to an old juniper woodland state (G), showing the duration of each state and the effects of wildfires. The duration of the successional stages and results of wildfires have been described by Bunting and others (2007). Landscape characteristics of sagebrush-steppe–juniper woodland mosaics under various modeled prescribed fire regimes.

In this approach the annual probability of a cell in one class to change to the following class is estimated as the inverse of the duration of the previous class. For example, if class A has a duration of 10 years, then each cell in that class has an annual probability of 1/10 = 0.10 to change to class B. The annual probabilities of wildfires are also different for the different classes. Annual probabilities of fire in the different successional stages were estimated for a baseline scenario based on the fire history in the Owyhee Mountains between 1973 and 2014 (Figure 10.11).

Finally, to start the simulations, the initial composition of the landscape provided represents a typical watershed in the area studied. A summary table (Table 10.1) shows the values used for the annual probabilities for each class to advance to the next stage or to be burned and go back to the initial stage, as well as the initial landscape composition. The corresponding relative transition matrix with annual probabilities is shown in Table 10.2.

In Table 10.2, for simplicity, the probability of the initial class to remain in the same class was only considered dependent on the time it takes for succession and the probability that the final class will remain in the same class is only dependent upon the probability of disturbance (as there are no other subsequent classes). In fact, old woodland develops slowly and juniper trees greater than 1000 years old are commonly found.

The transition matrix shown in Table 10.2 provides the baseline scenario "business as usual" that can be used to simulate future landscape composition at different time steps. We can now use the same approach to create new transition matrices with different probabilities to evaluate different scenarios.

Important issues for future landscapes can be made as "what if" questions. One such question is that of the possible influence of changing climate and other factors of global change on fire regimes. It is apparent from statistical data[29] that in recent decades the areas burned in wildfires have been increasing in the region (Figure 10.12).

We can now use our transitional model to evaluate the possible outcomes of different scenarios in the change of wildfire regimes. We can use the baseline scenario as the average probabilities for the period 1973–2014 and we can now simulate the changes in landscape composition by applying different scenarios of change.

"What if" wildfire probability increases in all classes by 100%, 200%, or 300%? We can term these scenarios as Wildfire × 2, Wildfire × 3, and Wildfire × 4, and use the same

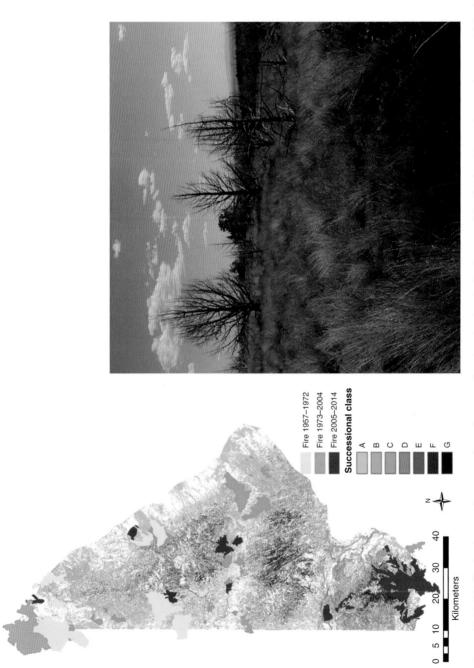

Figure 10.11 Wildfire has been an important factor for many decades, shaping the landscapes of the Owyhee Mountains, southwestern Idaho, USA. Areas burned in wildfires from 1957 to 2014 (left) and vegetation recovery 6 years after the Tongue–Crutcher wildfire of 2007 (right). Photo by Stephen Bunting.

Fire 1957–1972
Fire 1973–2004
Fire 2005–2014

Successional class

A
B
C
D
E
F
G

N

0 5 10 20 30 40
Kilometers

Table 10.1 Parameters for the simulation of landscape dynamics in the sagebrush steppe/juniper woodland succession. For each vegetation stage there is the indication of its proportion in the initial composition $p_j(0)$, its annual probability to burn (d_j), and the annual probability of changing to the next successional stage (e_j), which is inversely correlated to the duration of the stage.

Code	Vegetation successional stage (j)	Annual probability of changing to the next stage (e_j)	Annual probability of wildfire (1973–2014) (d_j)	Initial composition $V(0)$ $p_j(0)$
A	Grass forb herbland	0.100	0.004	0.00
B	Open sagebrush steppe	0.067	0.004	0.20
C	Closed sagebrush steppe	0.033	0.003	0.25
D	Initial woodland development	0.050	0.002	0.20
E	Mid woodland development	0.040	0.002	0.20
F	Advanced juniper woodland development	0.004	0.001	0.10
G	Old juniper woodland	0.002	0.002	0.05

Table 10.2 The relative transition matrix (M) resulting from the application of the simulation parameters in Table 10.1.

		To class						
From class		A	B	C	D	E	F	G
	A	0.900	0.100	0	0	0	0	0
	B	0.004	0.929	0.067	0	0	0	0
	C	0.003	0	0.964	0.033	0	0	0
	D	0.002	0	0	0.948	0.050	0	0
	E	0.002	0	0	0	0.958	0.040	0
	F	0.001	0	0	0	0	0.995	0.004
	G	0.002	0	0	0	0	0	0.998

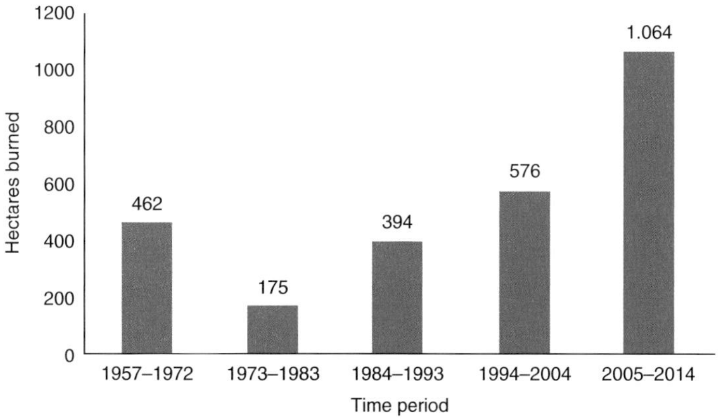

Figure 10.12 Area burned annually in the Owyhee Mountains by time period.

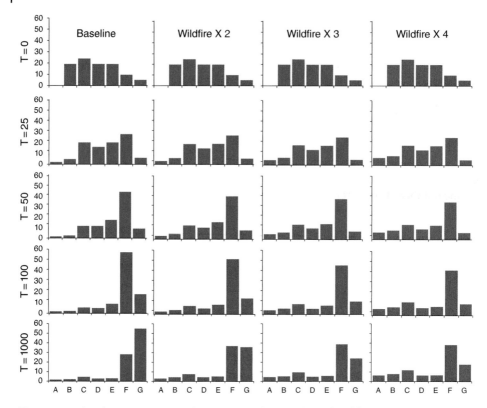

Figure 10.13 Landscape composition (in percent on the Y-axis) simulated for time steps (T = 0, 25, 50, 100, and 1000 years) for the four scenarios used in the simulations: baseline (average wildfire probabilities 1973–2014) and wildfire frequency for all classes multiplied by 2, 3, and 4.

approach multiplying by 2, 3, or 4 the annual probabilities of wildfire determined for the baseline scenario.

Even though it is known that there may be several problems associated with running ecological models for long time periods, there are also good reasons to justify this option. For example, Keane and others[30] recommend a simulation timespan long enough for the majority of the landscape units to experience at least 3–5 fires during the simulations. Therefore, we decided to run the models for 1000 years into the future as there are often long-term effects that may not be readily observed when modeling over shorter time periods such as 100 years. In addition it may be difficult to determine when the landscapes will theoretically achieve a steady-state shifting mosaic condition. The results of the simulations for the four scenarios are shown in Figure 10.13.

The results shown in Figure 10.13 indicate that, with the current low wildfire probabilities resulting from the current wildfire suppression strategy, the tendency of landscapes' composition in the Owyhee Mountains will continue to show an increasing dominance by juniper woodland (stages F and G) at the expense of sagebrush steppe (stages B and C). Our modeling output agreed with previous research that indicated that within 100 years our landscape would be dominated by advanced woodland (stage F). Our results indicate that, after 1000 years, the landscapes will tend to be dominated by the last stage (G, old juniper woodland). Major differences from that trend are only found after 1000 years when the proportion of class G is shown to decrease significantly with increased wildfire probability (especially in the scenario Wildfire × 4) allowing for a better representation of the classes of the seral stages.

An equivalent exercise can be made using alternative management strategies. A previous study[31] on this issue used a distributional dynamic landscape model, the vegetation dynamics development tool (VDDT)[32]. In that system, the simulated landscape is composed of cells, and the area of each cell and that of the landscape are determined by the user. VDDT moves cells from one vegetation composition to another via deterministic transitions based on time (years), which define the successional pathway. When a cell initially enters a stage along the pathway, it will remain in that composition for a specific number of years, referred to as time steps, unless a disturbance factor affects that cell. Disturbance factors, such as fire, insect and pathogen outbreaks, domestic animal grazing, and forest harvest, are represented by probabilistic transitions. Probabilities can range from 0 to 1, where a 0 means that the disturbance will not occur at all and a 1 means that the disturbance will occur at every time step. A probability of 0.1 means that the probability will occur every 10 time steps on average. The probabilistic transitions lead to predetermined stages along the successional pathway. The model is run for a specific number of time steps. The model output is not spatially explicit but does allow calculation of the landscape metrics that are based on composition, such as diversity, richness, and evenness.

We can use the same transition model to illustrate the efficacy of different management alternatives with the goal of addressing the loss of sagebrush steppe in this ecosystem. Recall that it is known that sagebrush steppe (Figure 10.14) is critical to many organisms and that it has been declining in this area for the past half century. Therefore it is considered that the management of the landscape should meet the goal of maintaining at least 25% of the landscape in sagebrush steppe (classes B and C).

Figure 10.14 Sagebrush steppe stages can be classified as open canopy: class B, background in the picture on the left; or closed canopy: class C, foreground in the left picture. Mature juniper woodlands can be classified as advanced juniper woodland (class F) or old juniper woodland (class G), picture on the right. Photos by Stephen Bunting.

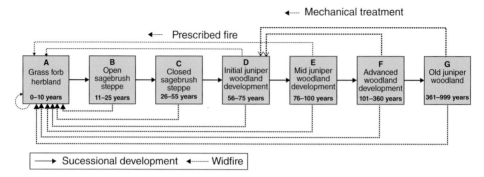

Figure 10.15 The transition model: the solid arrows indicate deterministic successional transitions and the dashed arrows indicate probabilistic disturbance transitions resulting from wildfire or management alternatives as prescribed fire or mechanical treatments.

Mature juniper woodland is also critical habitat for many organisms and requires 400–500 years to develop. It is also considered that the management of that landscape should meet the requirement of retaining at least 15% of the landscape in mature juniper woodland (classes F and G).

In order to assume the stability of the transition matrix (Markovian process) it was considered that, although climate varied between years, the mean climate did not change during the model period. The management treatments were superimposed on the baseline scenario.

Similarly to what was done before, four scenarios, now representing management alternatives, were used in the simulations. The diagrammatic picture of the different management options is presented in Figure 10.15, and images of the two types of management treatments (prescribed fire and mechanical) are shown in Figures 10.16 and 10.17.

The four management scenarios used in the simulations were:

1) The baseline scenario (as before only with average wildfire probabilities 1973–2014).
2) The prescribed burning scenario (baseline probabilities with increased probabilities of fire $d = 0.02$ in classes D and E that result in changes to class A).
3) The mechanical treatment scenario (baseline probabilities with mechanical treatment in class F with $d = 0.01$ and class G with $d = 0.005$, resulting in changes to class D).
4) A combination of prescribed burning and mechanical treatment (adding baseline probabilities to those associated with prescribed burning and mechanical treatments).

The successional stages and results of wild and prescribed fires have been described by Bunting and others[33] and the mechanical treatment was added for this illustration. The probabilities for prescribed burning and mechanical treatments were developed by the authors to best achieve the management goals. The initial landscape composition is the same as before. Table 10.3 summarizes the input data for the model.

The relative transition matrix with annual probabilities corresponding to scenario 4 is shown in Table 10.4. The four management scenarios were then used to make simulations of future landscapes for 1000 years. Results are shown in Figure 10.18.

Figure 10.16 September prescribed burn in western juniper woodland, southwestern Idaho. Photo by Stephen Bunting.

Figure 10.17 Recent vegetation response after mechanical treatment of western juniper (*Juniperus occidentalis*) using a mastication machine, which grinds up juniper trees into pieces 15 cm or less. This treatment is frequently used as a restoration technique to control the encroachment of juniper woodland to maintain sagebrush steppe vegetation. Photo by Stephen Bunting.

Table 10.3 Parameters for the simulation of landscape dynamics in the sagebrush-steppe juniper woodland succession including management alternatives. The values for the initial composition (p_j), annual probability of changing to the next successional stage (e_j), and annual probability to burn are the same as in Table 10.1 (reference scenario). Two new columns are added to the probability of wildfire to estimate the disturbance probability (d_j): these represent the probabilities associated with the management treatments (prescribed fire and mechanical treatment).

			Components of disturbance (d_j)		
Code	Vegetation successional stage	Annual probability of changing to the next stage (e_i)	Annual probability of wildfire (1973–2014)	Probability of prescribed fire (back to class A)	Probability of mechanical treatment (back to class D)
A	Grass forb herbland	0.100	0.004	0	0
B	Open sagebrush steppe	0.067	0.004	0	0
C	Closed sagebrush steppe	0.033	0.003	0	0
D	Initial woodland development	0.050	0.002	0.02	0
E	Mid woodland development	0.040	0.002	0.02	0
F	Advanced juniper woodland development	0.004	0.001	0	0.010
G	Old juniper woodland	0.002	0.002	0	0.005

Table 10.4 The relative transition matrix resulting from the application of the simulation parameters in Table 10.3 under scenario 4 (wildfire as baseline plus prescribed fire and mechanical treatment).

				To class			
From class		A	B	C	D	E	F	G
	A	0.900	0.100	0	0	0	0	0
	B	0.004	0.929	0.067	0	0	0	0
	C	0.003	0	0.964	0.033	0	0	0
	D	0.022	0	0	0.928	0.050	0	0
	E	0.022	0	0	0	0.938	0.040	0
	F	0.001	0	0	0.010	0	0.985	0.004
	G	0.002	0	0	0.005	0	0	0.993

In order to try to achieve our management goals we modeled a prescribed fire program (Scenario 2) that would be applied to vegetation stages D and E, as it has been shown that these stages are more easily burned than later woodland stages (F and G) because of greater amounts of herbaceous and smaller woody fuel components. In addition the vegetation responds more rapidly following a fire[34]. The annual proportion of burned area was set at 2%.

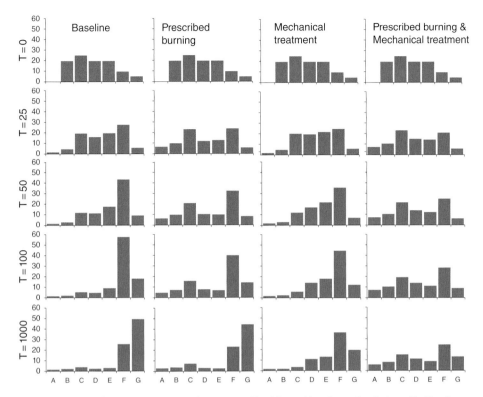

Figure 10.18 Landscape composition (percent on Y-axis) resulting from simulations (for T = 0 to 1000 years) for the four management scenarios used. Scenarios 1 to 4 from left to right.

The results indicate that under that prescribed fire program the sagebrush steppe would decline from its current levels (Scenario 2), but slower than in the wildfire only scenario (Scenario 1). After 100 years class F would dominate the landscape but after 1000 years it would be the final stage G that dominates. The management goal of maintaining an adequate proportion of sagebrush steppe is not achieved in the long run with this level of prescribed burning.

Scenario 3 includes more aggressive mechanical treatments of F and G. These treatments are not effective in retaining adequate levels of sagebrush steppe. After 1000 years it is young juniper woodland (F) that has a higher proportion in the landscape but old woodland (G) is also well represented. This management alternative maintains reasonable proportions of initial and mid woodland development stages (D and E) and mature juniper woodland (F and G). However, it is observed that, in the long run, the initial stages (A, B, and C) are virtually absent from the landscape. The management goal of maintaining adequate levels of sagebrush steppe (classes B and C) are not met in this scenario.

The last management scenario (4), with a combination of prescribed fire and mechanical treatments, results in a more rapid development of landscapes with a more diverse landscape composition, where all the classes are relatively well represented (high evenness). The various indices of diversity and evenness already explained in the previous chapters can be easily applied and confirm this conclusion.

It can be concluded that this management alternative meets the required management goals and rapidly results in a balanced landscape. Therefore the combination of prescribed burning in the younger successional stages and mechanical treatments in more developed woodlands seems to be the most interesting scenario.

This example illustrates the usefulness of the simple approach of transition models in evaluating alternative management scenarios through the simulation of the dynamics of future landscape composition.

These exercises are very useful in contributing to our understanding of the system dynamics. However, they should always be considered as models that simplify reality. Some of the basic assumptions of these models are generally not completely met, as in the assumption of transition matrices being constant through time, a requirement for the use of results originating from Markovian processes. Also, it is important to recall that scales matter. Spatial and temporal scales have to be considered. For a small landscape it is likely that one large wildfire event, for instance, will cover the entire landscape. In this case the behavior of a shifting mosaic where only a small proportion of the landscape is changing can only be observed in larger landscapes.

Relatively rare but broad scale events can result in major changes to large landscapes. For example, the combination of several years of extreme drought combined with pine beetle and other insect epidemics resulted in widespread mortality of native pines including ponderosa pine (*Pinus ponderosa*) and sugar pine (*Pinus lambertiana*) trees in the southern Sierra Mountains of California, USA (Figure 10.19). In 2017 it was estimated that more than one hundred million pine trees had died in California[35]. Pine trees of all size classes were affected. Many trees killed were hundreds of years old. Rare events such as these are nearly impossible to predict in landscape modeling frameworks given their low probability and the brief temporal scale of our data. However, these effects will influence the Sierra Nevada Mountains landscape composition and processes for decades if not hundreds of years into the future.

Despite the indicated limitations of assumptions and of scale, the approach of the transition matrices has been very useful in evaluating landscape changes and planning landscape management by comparing different alternative scenarios.

10.5 Combining Spatial and Temporal Analysis in Transition Models

As described above, transition models are very useful for understanding landscape dynamics and Markov chain processes, and they can be used to project future changes in landscape composition based on the immediately preceding state described in a transition probability matrix. However, this type of analysis does not provide any information about the spatial distribution of classes in the landscape, that is, there is no spatial component in the modeling outcome. Therefore, landscape composition can be modeled with transition matrices, but this approach does not provide information about landscape configuration.

In many cases, as in forest management, the spatial units are already defined in planning, and the management actions are applied to those spatial units. Typically, the size of these units is limited as extensive tree harvesting can have effects on soil protection,

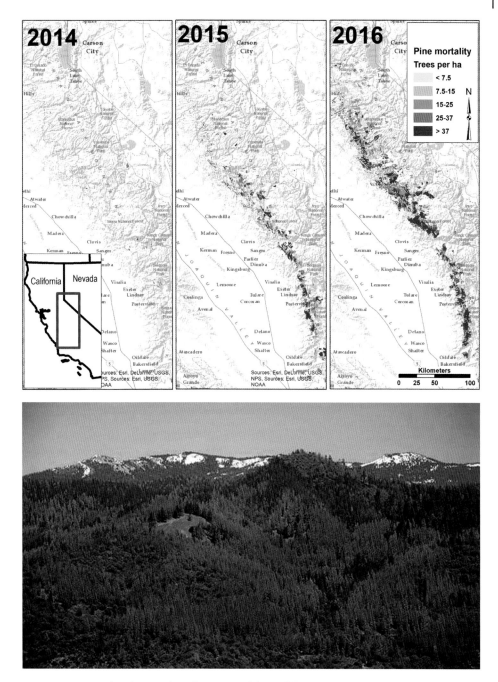

Figure 10.19 Broad scale mortality of pines in California following rare and extreme drought periods and extensive beetle attacks. Maps (above) and photo (below) showing the very large extent of the damage over a short time period. *Sources*: Data source, https://www.fs.usda.gov/detail/r5/forest-grasslandhealth/?cid=fsbdev3_046696, image by Eva K. Strand (above); https://www.nasa.gov/feature/ames/aerial-images-show-decades-of-foothill-forest-growth-erased-due-to-california-s-extreme (below).

Figure 10.20 Forest pattern resulting from wood harvesting in western Washington, USA. *Source:* Google Maps.

biodiversity, and natural regeneration of the forests. In addition to achieving sustained and continuous output of products the harvest unit size may also be determined by the total area of forest managed.

Within this framework the transition matrices presented earlier are fundamental to decide on the proportion of the mature forest to be harvested in each year (Figure 10.20), but the spatial allocation of the harvesting operations is decided by the manager in another step.

However, in the absence of predefined planned spatial units, the map cells are the spatial units used in the analysis and the modeling. In this case, if there is no spatial information in the transition matrices presented, all cells (or patches) of the same class will have the same probability to change to another class irrespective of its location. This would occur in a random neutral process where classes change randomly in the landscape and no interactions between a cell and nearby cells occur. However, as we saw before in real landscapes represented by cells of different classes, cells of the same type tend to occur more frequently in the proximity of similar cells. This can occur because of existing processes of contagion promoting expansion of that class (e.g., seed regeneration from forests creating new forests in the proximity), because of external preexisting factors (e.g., a certain class can only occur in soils of a certain parent material or on a given slope class), because of physical factors associated with the process (e.g., a certain change can only occur if a certain disturbance occurs), or because of social factors (e.g., certain changes can be limited in some areas because of legal restrictions).

In order to add spatial dependence to landscape change modeling, we need to be able to differentiate transition probabilities between cells (or patches) of the same class that are in different locations in the landscape. This can also be done with the use of the concept of conditional probabilities.

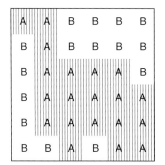
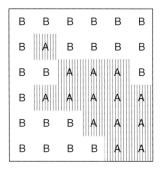

Global Transition Matrix

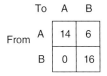

To	A	B
From A	14	6
B	0	16

Relative Transition Matrix

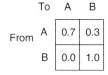

To	A	B
From A	0.7	0.3
B	0.0	1.0

Figure 10.21 Two different representations with two classes (A and B) corresponding to two different moments in time of the same landscape (at time t: $p_A = 0.56$, $p_B = 0.44$; at time t+1: $p_A = 0.39$, $p_B = 0.61$). A global or absolute transition matrix (AT) is computed with the absolute values (number of cells) of the transitions as well as a relative transition matrix (M) with the proportions of a class in the first time that changed to a class in the second time.

A first example can illustrate the concept of spatially explicit conditional probabilities. We can use a simple two-class landscape and compute, as before, the transition probabilities between the two classes A and B (Figure 10.21).

In the case shown in Figure 10.21 all cells in class B will have a probability of 1 to remain in class B and zero probability to change to class A. In any simulation all cells already in class B are expected to remain in the same class. For cells that are in class A, the probability of remaining in class A is 0.7 and the probability of changing to class B is 0.3. Then, in the simulation of the landscape at the next period, the allocation of any cell to the future class can be performed by generating a random value (between 0.0 and 1.0 from a uniform distribution) and comparing it to the cumulative probabilities of the outcomes. In this case we have, for a cell that is class A:

Cumulative probability of a cell in class A remaining in the same class: from 0.00 to 0.70
Cumulative probability of a cell in class A to change to class B: from 0.70 to 1.00

Then, for any given cell in class A, if the random value falls between 0.00 and 0.70 the cell is expected to remain in class A; if the random value falls between >0.70 and 1.00 the cell is expected to change to class B. Table 10.5 illustrates the procedure.

Now we can simulate the landscape dynamics in the following period. From the 14 cells of class A in the second map, 10 cells are predicted to remain in class A in the next period as the random value is below the threshold 0.7. The other 4 cells, where the random value was above 0.7, were predicted to change to class B. The outcome (one of many possible maps representing the landscape composition and configuration in the next period) is shown in Figure 10.22.

Table 10.5 Procedure to illustrate the assignment of a new class to the cells belonging to class A in the first moment. The first column is a sequence of randomly generated numbers from 0 to 1. These values are compared with the threshold 0.7 defined as the global cumulative probability of a cell in class A to remain in the same class. If the random value exceeds 0.7 then the cell is expected to change from A to B.

Random value	Comparison	Change from class A to Class
0.614	<0.7	A
0.638	<0.7	A
0.768	>0.7	B
0.161	<0.7	A
0.584	<0.7	A
0.237	<0.7	A
0.606	<0.7	A
0.963	>0.7	B
0.636	<0.7	A
0.623	<0.7	A
0.702	>0.7	B
0.541	<0.7	A
0.554	<0.7	A
0.722	>0.7	B

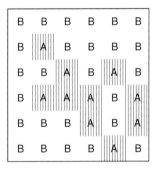

Figure 10.22 A possible outcome of a landscape in the sequence of those presented in Figure 10.21 using a simulation with the global transition matrices and the results of Table 10.5. All cells of class A had the same probability of remaining in class A (0.7) or changing to class B (0.3) regardless of their location in the landscape.

This exercise shows how a simulation of future landscapes could be generated with no additional information about the spatial location of the cells.

However, we can look at the same example using an auxiliary map of a certain variable that is known or supposed to have an influence on the transitions between the two classes. As an example we will consider that the variable in question is slope and that we can subdivide our landscape in two categories, the flat terrains (≤5%) and the slopes (>5%). In this case two different relative transition matrices could be computed: one for areas with a flat terrain and another for slopes. Figure 10.23 shows an example to illustrate the procedure.

We now have two different transition matrices applying to different areas of the map. We can also see that some transition probabilities for cells of the same class can be considered to be different, depending on their spatial location. For example, the probability of a cell to change from class A to B is 0.6 on flat terrain whereas the corresponding probability on slopes is 0.0. We can see this situation as expressed as spatially explicit

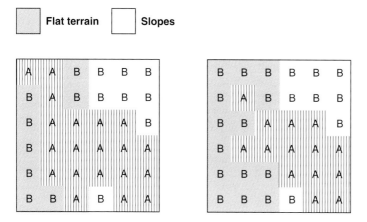

For flat terrain

Absolute Transition Matrix

	To	A	B
From	A	4	6
	B	0	8

Relative Transition Matrix

	To	A	B
From	A	0.4	0.6
	B	0.0	1.0

For slopes

Absolute Transition Matrix

	To	A	B
From	A	8	0
	B	0	8

Relative Transition Matrix

	To	A	B
From	A	1.0	0.0
	B	0.0	1.0

Figure 10.23 The same situation of landscape dynamics shown in Figure 10.21 but considering that slopes have an effect on transitions. The left side of the landscape is on flat terrain whereas the right side of the landscape is on slopes. Absolute and relative transition matrices are computed separately for the two conditions. Class B remains stable in both conditions. Class A remains stable only in slopes.

conditional probabilities where the transition probabilities between classes are conditional to the topography (flat terrains or slopes).

The changes in the cells of class A in the landscape can now be simulated and results are presented in Table 10.6, using for comparison the same random values as in Table 10.5.

It is clear from Table 10.6 that cells of class A in flat terrain are likely to change to class B (3 of 4 cells change and only the last one remains unchanged) whereas for slopes all 10 cells of class A remain in the same class. The outcome of this simulation results in the map of Figure 10.24.

Table 10.6 Procedure to illustrate a simulation for the assignment of a new class to the cells belonging to class A in the first moment. In this case the landscape is subdivided into two areas (flat terrain and slopes) with different thresholds for the comparison with the random values. For flat areas a random value above 0.4 would mean a prediction to change to class B whereas all cells of class A in slopes were predicted to remain in the same class.

Condition	Random value	Comparison	Change from class A to Class
Flat terrain	0.614	>0.4	B
	0.638	>0.4	B
	0.768	>0.4	B
	0.161	<0.4	A
Slopes	0.584	<0.1	A
	0.237	<0.1	A
	0.606	<0.1	A
	0.963	<0.1	A
	0.636	<0.1	A
	0.623	<0.1	A
	0.702	<0.1	A
	0.541	<0.1	A
	0.554	<0.1	A
	0.722	<0.1	A

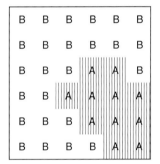

Figure 10.24 A simulated landscape in the sequence of those presented in Figure 10.21 using a spatially explicit conditional transition and the results of Table 10.6. Cells of class A in flat terrain are likely to change to class B but those on slopes remain in class A. In all conditions cells in class B remain unchanged.

By simulating maps of projected changes using conditional probabilities, new cells of class B would only occur on flat terrain and none would be created on slopes. Given the apparent importance of slopes in the process the results of this simulation seem to be more realistic than those obtained earlier without considering slope in the process.

The use of spatially explicit conditional probabilities as a tool to develop more realistic maps of possible future landscapes is not restricted, however, to preexisting factors. As we saw in the preceding chapter, the occurrence of disturbances, such as wildfires, can also be considered an important conditional factor for the transition probabilities of the various landscape classes and can be used in the simulations of future landscapes. In our example we could define the left part of the landscape, considered to be a flat terrain, as representing an area that burned and that changes from class A to class B could result from that disturbance. We could also define the right part of the landscape as given a protection status where no further urban developments were allowed. If class A was an urban area then its development would not occur in the right part of the landscape. These are all examples of conditional transition probabilities arising from external factors that are not included in maps of land classes.

Using only the information provided by the maps of the landscape classes, we can also use the concept of conditional transition probabilities using proximity analysis. One particularly important case of the use of conditional probabilities to deal with spatial proximity is in the use of cellular automata. In the simplest case where the whole landscape is equally suitable for all the changes, these changes may be only a function of the proximity of existing cells of the same class. A patch of cells of a given class will then grow as a contagion process from the existing boundary. In practice, this hybrid spatial and temporal system based on cellular automata and Markov chain analysis has been used in a module known as CA_MARKOV in the GIS analysis tool IDRISI[36]. A contiguity filter is applied as a spatially explicit weighting factor for the transition probabilities for a given class, weighing more heavily those that are in the proximity of existing cells of the same class. Different contiguity filters using different neighborhood methods can be used (Figure 10.25).

In general, simple contagion models using contiguity filters and proximity analysis are very useful to understand and explain certain phenomena, but in reality landscape changes are often more complex. Also, simple models that explain the changes only by suitability or constraints without taking into account the spatial proximity of cells of the same class are often too limited to adequately represent the reality of landscape changes.

Therefore, in many cases, it is convenient to use both contiguity filters combined with suitability maps to indicate the areas where changes are more likely to occur due to criteria such as proximity to existing classes or to roads, to water bodies, to areas with high population density, or other physical or social factors. A multicriteria evaluation can be used to create specific suitability land use/land cover maps based on the rules that associate these factors with the classes and their dynamics. These suitability maps are then used to assist the spatial allocation of the transitions.

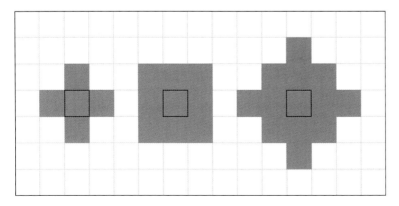

Figure 10.25 Typical configurations used in cellular automata processes based on different ways to define neighborhoods: von Neumann 3 × 3 (left), Moore 3 × 3 (center), and von Neumann 5 × 5 (right).

Combinations of spatial and temporal analyses can then benefit from information of physical and social attributes of the landscape. These systems have been applied to various situations resulting in maps that may provide descriptions of possible future landscapes that are in agreement with the processes taken into account. In most cases the importance of the factors is not as strong as was illustrated by our example of dichotomous outcomes (flat terrain versus slopes). In these cases we can use gradients as suitability factors and create suitability maps that indicate the areas where changes are more likely to occur due to physical or social factors.

A multicriteria evaluation can then be used to create specific suitability land use/land cover maps based on the rules that associate these factors with the classes and their dynamics. These suitability maps are then used to assist the spatial allocation of the transitions. Also constraints such as those derived from planning laws can be included as factors. This was the approach initially proposed by J.R. Eastman in the development of the Land Change Modeler, included in the software IDRISI in 2006[37].

This system has been applied to various situations resulting in maps that provide descriptions of possible future landscapes that account for numerous processes and other restrictions. This work has continued with the extension of the Land Change Modeler for ArcGIS in 2015 and with the simultaneous development of Modules for Land Use Change Simulations (MOLUSCE) for QGIS.

An example of the land use dynamics of the Portuguese mainland can help to visualize the projected landscape changes (Figure 10.26). This work shows the composition and configuration of the Portuguese landscapes under a scenario of the continuation of the trends observed between 1990 and 2005, including urban growth in coastal areas and replacement of pines by eucalypts in central Portugal.

One major issue for the realistic behavior of the simulations is in the appropriate choice and combination of factors to generate suitability maps. A possible procedure to create suitability maps can be illustrated by the development of a suitability map for urban development for the metropolitan area of Seoul (South Korea)[38] (Figure 10.27).

In the next step it is important to consider the legal limitations for urban development. The final result for the analysis of the suitability areas for urban development around Seoul is shown in Figure 10.28.

The example of the definition of suitability classes for urban development in Seoul (Figure 10.29) includes the combination of physical and social factors that should be taken into account when analyzing and predicting landscape changes. From the figures it is apparent that suitability for urban growth tends to be in the proximity of roads, expressways, trains, or subway stations and in the vicinity of already existing urban areas, as in a contagion process.

The two previous examples (for Portugal and Seoul) illustrate the use of suitability maps together with Markov analysis and cellular automata in assisting with forecasts of landscape changes.

However, there are situations where physical or legal constraints are more important in the process of landscape change, there are others where land physical suitability is more relevant, and still others where the contagion processes are more influential.

When major landscape processes are contagious by nature the use of contiguity filters associated with transition matrices is particularly helpful in modeling. It is therefore easy to understand that many successful examples of this approach are found in the

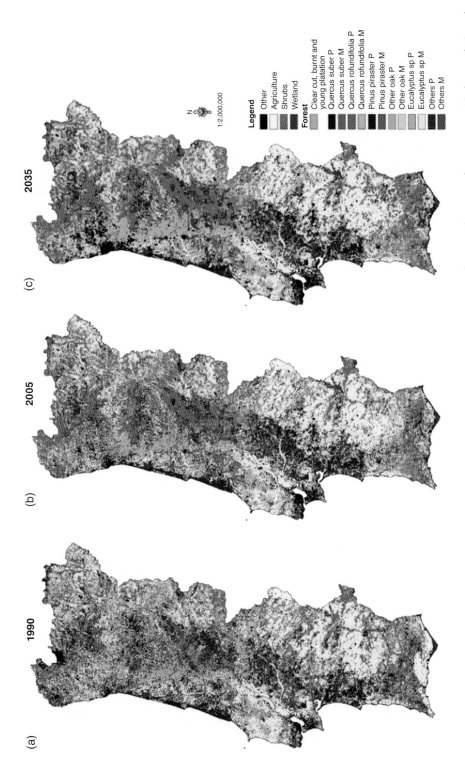

Figure 10.26 Observed landscape changes between 1990 and 2005 in the Portuguese mainland and a forecasted scenario for 2035 using Markov analysis and cellular automata. *Source:* Godinho-Ferreira, P., Magalhães, M., Tomé, M., and Rego, F.C. Unpublished data. Portuguese Forest Dynamics. Future scenarios and implications for forest management and policy decisions. Presentation at the IUFRO Conference on *Mixed and Pure Forests in a Changing World 2010*, Vila Real, Portugal.

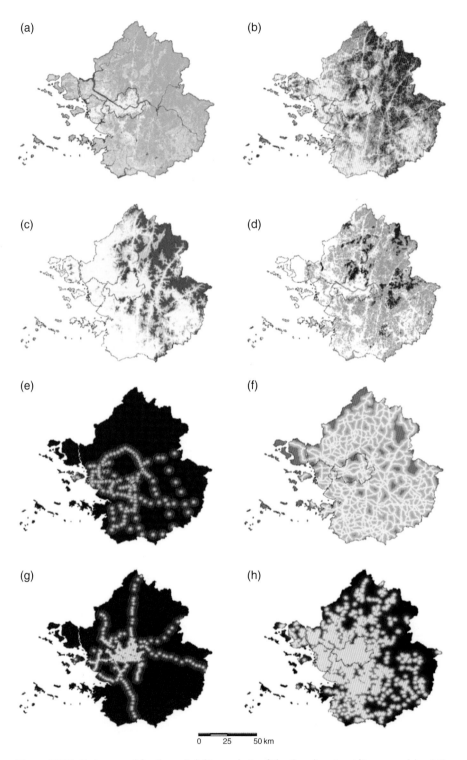

Figure 10.27 Factors used for the suitability analysis of the Seoul metropolitan area: (a) existing land use, (b) slope, (c) altitude, (d) ecoregion, (e) distance from expressway gate, (f) distance from main road, (g) distance from train or subway station, and (h) distance from existing urban area. *Source:* Adapted from Choi, H.S. and Lee, G.S. (2016) Planning support systems (PSS)-based spatial plan alternatives and environmental assessment. *Sustainability*, 6, 286.

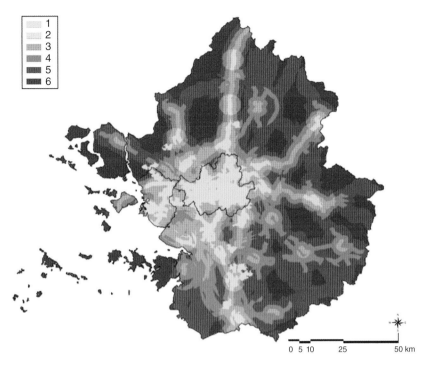

	1
	2
	3
	4
	5
	6

0 5 10   25   50 km

Figure 10.28 Final map of suitability classes (1-6) for urban development near Seoul. *Source:* Choi, H.S. and Lee, G.S. (2016). Planning support systems (PSS)-based spatial plan alternatives and environmental assessment. *Sustainability*, 6, 286.

Figure 10.29 An image of Seoul, South Korea. *Source:* https://commons.wikimedia.org/wiki/File: Seoul_Tower_View.jpg.

literature associated with contagion landscape processes, for example urban sprawl, as illustrated in studies in various areas of the world from the United States[39], Australia[40], and Iran[41].

A popular model that uses the cellular automaton approach for the computational simulation of urban growth and land use changes that are caused by urbanization is SLEUTH. This "detective" model originated in 1993 from work on another contagion process, the propagation of wildfires[42]. The initial application of SLEUTH for urban growth was focused on the San Francisco Bay area[43] in 1997 and it has been used in many other cities in the United States since then.

The first application of the SLEUTH model outside the United States was conducted in Portugal, in the Metropolitan areas of Lisbon and Porto[44], and has been applied in many other cities in the world, especially in rapidly growing cities in fast developing countries such as China.

We will use the example of Changsha[45], China, to illustrate the process. The first steps in the analysis were to determine the best coefficients to use for transitions by model calibration and validation. After these steps, the model was used to simulate the observed changes in the validation period. Yin and others simulated changes in the landscape around the city of Changsha between 1999 and 2005 and compared them to the actual changes. In spite of some observed differences, especially around the road networks, an importance in the urbanization process that was overemphasized by the model, the simulated and actual landscapes were quite similar.

The validated model was then used to simulate the future urban growth under different scenarios. The planning scenarios used were:

Scenario 1. Urban areas grow according to the current trend with no consideration of loss of farmland, forestland, and other green areas.
Scenario 2. Aimed to protect farmland, urban growth occupies the least farmland, forestland, and other green areas with no consideration of current trends.

The results of the simulations for 2015 with the two different scenarios can be compared with the actual landscape image in 2017 (Figure 10.30).

This comparison shows that the projection provided by Scenario 2, where protection of farmland, forests, and green areas was included, was much closer to reality than that of Scenario 1, where the trend observed before 2005 was predicted to continue.

It can easily be concluded from the observation of the images that the explosion of urban growth predicted in Scenario 1 did not occur, possibly because the conservation measures included in Scenario 2 were largely effective. In fact, in 2015 the city of Changsha has been awarded the "2015 China Sustainable City" for achieving high human development while also minimizing damage to the environment (Figure 10.31).

This example shows that we can use the concepts of Landscape Ecology to predict the future landscape composition and configuration also in the context of urban development. We can also use these tools to simulate future scenarios and to evaluate the consequences of legal restrictions of landscape changes and different land use policy and planning alternatives.

In summary, we have seen that, in many different contexts from true wilderness areas to landscapes dominated by urban development, the principles and techniques of Landscape Ecology are useful to inform Landscape Management. This is, after all, a primary objective of Applied Landscape Ecology.

Figure 10.30 Simulation results for the urban growth of Changsha in 2015 with the two planning scenarios (see the text for details) compared with the actual landscape. *Sources*: Google Earth 2017, and data from Yin, C., Yu, D., Zhang, H., *et al.* (2008) Simulation of urban growth using a cellular automata-based model in a developing nation's region, in *Geoinformatics and Joint Conference on GIS and Built Environment. Sixth International Conference on Advanced Optical Materials and Devices* (eds L. Liu, X. Li, K. Liu, *et al.*), International Society for Optics and Photonics.

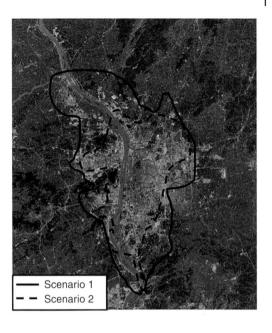

Scenario 1
Scenario 2

Figure 10.31 The Nianjia Lake in the center of city of Changsha in September 2012. *Source:* Yinsanhen own work [CC BY-SA 4.0 (https://creativecommons.org/licenses/by-sa/4.0)], from Wikimedia Commons.

Key Points

- Understanding how natural processes influence landscape dynamics is an important aspect of applied landscape management.
- The use of transition matrices are effective in simulating changes in disturbance regimes and predicting future landscape composition resulting from management activities.
- The use of transition matrices are an effective means to model landscape dynamics to meet management objectives.
- To add spatial character to landscape change modeling, we need to differentiate transition probabilities between cells (or patches) of the same class that are in different locations in the landscape. This can be done with the use of conditional probabilities. Contagion can be modeled by assigning different conditional probabilities to adjacent and faraway pixels or polygons.
- The use of landscape models, both spatially specific and distributional models, are important tools that land planners and land managers can utilize to incorporate the principles of Landscape Ecology into applied landscape management.

Endnotes

1 The ProSilva Association (1997) The Pro Silva Declaration of Apeldoorn 1997, https://prosilvaeurope.wordpress.com/prosilva-declaration-of-apeldoorn-1997/ (Accessed Mach 21, 2017).

2 McArthur, A.G. (1962) *Control Burning in Eucalypt Forests*, Commonwealth of Australia, Forestry and Timber Bureau, Forest Research Institute, Leaflet 80, Canberra, ACT.

3 Sneeuwjagt, R.J. and Peet, G.B. (1979) *Forest Fire Behaviour Tables for Western Australia*, Forests Department, Western Australia.

4 Mount, A.B. (1983) The case for fuel management in dry forests, Paper to Research Working Group No. 6 on Fire Research, Hobart, Tasmania.

5 Sneeuwjagt, R. (2008) *Prescribed burning: How effective is it in the control of large forest fires?* Department of Environment and Conservation Report WIT.135.001.0099, Perth, WA, Australia.

6 Abbot,t I., VanHeurck, P., and Burbidge, T. (1993) Ecology of the pest insect jarrah leaf miner (*Lepidoptera*) in relation to fire and timber harvesting in Jarrah Forest in Western Australia. *Australian Forestry*, 56, 264–275.

7 Lang, S. (1997) *Burning the Bush: A Spatio-Temporal Analysis of Jarrah Forest Fire Regimes*. Honours Thesis (Geography), Australian National University (ANU), Canberra.

8 Abbott, I., Van Heurck, P., and Burbidge, T. (1993) Ecology of the pest insect jarrah leaf miner (*Lepidoptera*) in relation to fire and timber harvesting in Jarrah Forest in Western Australia. *Australian Forestry*, 56, 264–275.

9 Lang, S. (1997) *Burning the Bush: A Spatio-Temporal Analysis of Jarrah Forest Fire Regimes*. Honours Thesis (Geography), Australian National University (ANU), Canberra.

10 Usher, M.B. (1966) A matrix approach to the management of renewable resources, with special reference to selection forests. *Journal of Applied Ecology*, 3, 355–367.

11 Usher, M.B. (1969) A matrix model for forest management. *Biometrics*, 25, 309–315.

12 Horn, H.S. (1975) Markovian properties of forest succession, in *Ecology and Evolution of Communities* (eds M.L. Cody and J.M. Diamond), The Belknap Press of Harvard University Press, Cambridge, MA, and London, England.

13 Horn, H.S. (1975) Markovian properties of forest succession, in *Ecology and Evolution of Communities* (eds M.L. Cody and J.M. Diamond), The Belknap Press of Harvard University Press, Cambridge, MA, and London, England.

14 Grime, J.P. (1973) Control of species density in herbaceous vegetation. *Journal of Environmental Management*, 1, 151–167.

15 Connell, J.H. (1978) Diversity in tropical rain forests and coral reefs. *Science*, 199, 1302–1310.

16 Leslie, P.H. (1945) On the use of matrices in certain population mathematics. *Biometrika*, 33, 183–212.

17 Usher, M.B. (1966) A matrix approach to the management of renewable resources, with special reference to selection forests. *Journal of Applied Ecology*, 3, 355–367.

18 Buongiorno, J. and Michie, B.R. (1980) A matrix model of uneven-aged forest management. *Forest Science*, 26, 609–625.

19 Sallnäs, O. (1990) A matrix growth model of the Swedish forest. *Studia Forestalia Suecica*, no. 183.

20 Schelhaas, M.J., Varis, S., Schuck, A., and Nabuurs, G.J. (1999) *EFISCEN's European Forest Resource*, European Forest Institute, Joensuu, Finland.

21 Verkerk, P.J., Schelhaas, M.J., Immonen, V., *et al.* (2016) *Manual for the European Forest Information Scenario model (EFISCEN 4.1)*, European Forest Institute Technical Report 99.

22 Sallnäs, O., Berger, A., Räty, M., and Trubins, R. (2015) An area-based matrix model for uneven-aged forests. *Forests*, 6, 1500–1515.

23 Packalen, T., Sallnäs, O., Sirkia, S., *et al.* (2014) *The European Forestry Dynamics Model: Concept, Design and Results of First Case Studies*, Publications Office of the European Union, EUR 27004.

24 Buongiorno, J., Peyron, J.L., Houllier, F., and Bruciamacchie, M. (1995) Growth and management of mixed-species, uneven-aged forests in the French Jura: Implications for economic returns and tree diversity. *Forest Science*, 40, 429–451.

25 Ingram, D. and Buongiorno, J. (1996) Income and diversity trade-off from management of mixed lowland dipterocarpa in Malaysia. *Journal of Tropical Forest Science*, 9, 242–270.

26 Boscolo, M., Buongiorno, J., and Panayotou, T. (1997) Simulation options for carbon sequestration through improved management of a lowland tropical rainforest. *Environment and Development Economics*, 2, 241–263.

27 Spathelf, P. and Durlo, M.A. (2001) Transition matrix for modeling the dynamics of a subtropical seminatural forest in southern Brazil. *Forest Ecology and Management*, 151, 139–149.

28 Liang, J. and Picard, N. (2013) Matrix model of forest dynamics: An overview and outlook. *Forest Science*, 59, 359–378.

29 US Department of the Interior, Bureau of Land Management (BLM), Idaho State Office, Fire Perimeters Historic, vector data; http://inside.uidaho.edu/.

30 Keane, R.E., Cary, G.J., and Parsons, R. (2003) Using simulations to map fire regimes: An evaluation of approaches, strategies, and limitations. *International Journal of Wildland Fire*, 12, 309–322.

31 Bunting, S.C., Strand, E.K., and Kingery, J.L. (2007). Landscape characteristics of sagebrush-steppe–juniper woodland mosaics under various modeled prescribed fire regimes, in *Proceedings of 23rd Tall Timbers Fire Ecology Conference: Fire in Grassland and Shrubland Ecosystems* (eds R.E. Masters and K.E.M. Galley), Tall Timbers Research Station, Tallahassee, FL.

32 ESSA Technologies Ltd (2007) *Vegetation Dynamics Development Tool User Guide*, Version 6.0. Prepared by ESSA Technologies Ltd, Vancouver, BC, http://essa.com/wp-content/uploads/2013/10/VDDT-60-User-Guide.pdf (Accessed October 5, 2016).

33 Bunting, S.C., Strand, E.K., and Kingery, J.L. (2007). Landscape characteristics of sagebrush-steppe–juniper woodland mosaics under various modeled prescribed fire regimes, in *Proceedings of 23rd Tall Timbers Fire Ecology Conference: Fire in Grassland and Shrubland Ecosystems* (eds R.E. Masters and K.E.M. Galley), Tall Timbers Research Station, Tallahassee, FL.

34 Miller, R.F., Bates, J.D., Svejcar, T.J., *et al.* (2005) *Biology, Ecology, and Management of Western Juniper*, Oregon State University, Agricultural Experiment Station, Technical Bulletin 152.

35 USDA Forest Service Pacific Southwest Research Station, https://www.fs.fed.us/psw/topics/tree_mortality/california/index.shtml (Accessed January 19, 2018).

36 Eastman, J.R. (2006) *IDRISI Andes Tutorial*, Clark Labs, Clark University, Worcester, MA.

37 Eastman, J.R. (2006) *IDRISI Andes Tutorial*, Clark Labs, Clark University, Worcester, MA.

38 Choi, H.S. and Lee, G.S. (2016) Planning support systems (PSS)-based spatial plan alternatives and environmental assessment. *Sustainability*, 6, 286.

39 Sayemuzzaman, M. and Jha, M., (2014) Modeling of future land cover land use change in North Carolina using Markov chain and cellular automata model. *American Journal of Engineering and Applied Sciences*, 7, 292–303.

40 Lahti, J. (2008). *Modelling Urban Growth Using Cellular Automata: A Case Study of Sydney, Australia*, Master's Thesis submitted to the International Institute for Geo-Information Science and Earth Observation, Enschede, The Netherlands.

41 Nouri, J., Gharagozlou, A. Arjmandi, R., *et al.* (2014) Predicting urban land use changes using a CA–Markov model. *Arabian Journal for Science and Engineering*, 39.

42 Clarke, K.C., Olsen, G., and Brass, J.A. (1993) Refining a cellular automaton model of wildfire propagation and extinction, in *Proceedings of the Second International Conference on the Integration of Geographic Information Systems and Environmental Modeling*, Breckenridge, CO.

43 Clarke, K.C., Hoppen, S., and Gaydos, L. (1997) A self-modifying cellular automaton model of historical urbanization in the San Francisco Bay area. *Environment and Planning B*, 24, 247–261.

44 Silva, E.A. and Clarke, K.C. (2002) Calibration of the SLEUTH urban growth model for Lisbon and Porto, Portugal. *Computers, Environment and Urban Systems*, 26, 525–552.

45 Yin, C., Yu, D., Zhang, H., *et al.* (2008) Simulation of urban growth using a cellular automata-based model in a developing nation's region, in *Geoinformatics and Joint Conference on GIS and Built Environment. Sixth International Conference on Advanced Optical Materials and Devices* (eds L. Liu, X. Li, K. Liu, *et al.*), International Society for Optics and Photonics.

Appendix A

Description of Notation Used in Formulae and Metrics

	Description
Subscripts of variables	
i, j	individual element (point, quadrat, line, or patch) or individual measurement
j, k	class of similar elements (species or patch types)
a	value of the Hill series
x	number of points
v	condition of a transition process
General symbols	
α, β, γ	parameters or coefficients of equations
x, y	variables
n	number of landscape elements, individuals, or measurements considered in the analysis
t	time
m	number of classes (patch types) in the landscape
S	number of species in the area considered
X_i, Y_i	Cartesian coordinates of point i
d_i	distance from i to its nearest neighbor
d_{ij}	distance between i and j
A	area
L	length of a line or of a patch
P	perimeter of a patch
p	probability of a certain event
r	radius of a circle
p_j	proportion of class j
p_{jk}	proportion of class j in the first group in class k in the second group

(Continued)

Applied Landscape Ecology, First Edition. Francisco Castro Rego, Stephen C. Bunting, Eva Kristina Strand and Paulo Godinho-Ferreira.
© 2019 John Wiley & Sons Ltd. Published 2019 by John Wiley & Sons Ltd.

Population statistics

N	total population size
N_j	size of population of species j
N_{jk}	size of population of species j in habitat k
$N._k$	size of population of all species in habitat k
K	carrying capacity
T_e	time for local extinction
ε	per capita birth rate
μ	per capita mortality rate
σ	colonization rate
ϕ	extinction rate
p_{eq}	proportion of equilibrium

General statistics

MEAN	average value of the variable in the population observed MEAN $= \sum x_i / n$
VARIANCE	variance of the population observed VARIANCE $= \sum (x_i - \text{MEAN})^2 / n$
SE	standard error
$E(x)$	expected value of variable x
z	statistic of the normal distribution
χ^2	chi-square statistic

Metrics		Application context	Related variables	
Line/perimeter metrics				
l_i	length of line i	Line/Perimeter		
L_{jk}	total length of edge lines between classes j and k	Class	TLA	total length of auxiliary lines
TL	total length of lines (as edgelines) in the landscape	Landscape		
D	fractal dimension	Line/Perimeter/ Class/Landscape	s	size of a ruler or the side of a quadrat used to measure a line, or the size of segments used in auxiliary lines to define crossings with existing lines

Area metrics

a_i	area of patch i	Patch		
A_j	total area of class j in the landscape $A_j = \sum a_{ij}$	Class	a_{ij}	area of patch i of class j
MPA	mean patch area	Landscape	ACV	coefficient of variation of patch areas
TA	total area of the landscape	Landscape	A'	area in different measurement units

Density

λ	density or number of elements (points) per unit area	Landscape	
λ_l	line density (total length of lines per unit area)	Landscape	
λ_p	patch density	Landscape	
NP	number of patches	Landscape	

Shape metrics

LWRATIO	length to width ratio of a patch	Patch		
SHAPE	shape index of a patch	Patch	P_c	perimeter of a circle with area A
SHAPE2	a second shape index of a patch	Patch		
EDGE	the area of the patch within a certain distance from its perimeter	Patch		
CORE	the area of the interior of the patch at more than a certain distance from its perimeter	Patch		
CAI	core area index	Patch		
MSI	mean shape index	Class/Landscape		
AWMSI	area weighted mean shape index	Class/Landscape		
MEDGE	mean edge area	Class/Landscape	EDGECV	coefficient of variation of edge areas
MCORE	mean core area	Class/Landscape	CORECV	coefficient of variation of core areas
MCAI	mean core area index	Class/Landscape		
IEDGERATIO	interior (core) to edge ratio	Landscape		
LSHAPE	landscape shape index	Landscape		

(Continued)

Metrics		Application context	Related variables	
Pattern metrics				
PATTERN1	index of the spatial pattern of distribution of points	Class	n_x	number of quadrats with x points
PATTERN2	index of the spatial pattern of distribution of lines	Class	n'_x	expected number of quadrats with x points with a random distribution
PATTERN3	index of the spatial pattern of distribution of patches	Class	CROSS	number of crossings between lines and auxiliary lines
VMR	variance to mean ratio of the distribution of landscape elements	Class	$L(r)$	radius of an equivalent circle
$f(r)$	metric of point pattern for the Ripley's K analysis	Class	Q	number of quadrats
q_{jj}	proportion of like adjacencies for class j	Class	x_i	number of points in quadrat i
$CONTAG_j$	contagion index for class j	Class	a	area of quadrat
			EMD	expected mean nearest neighbor distance between elements
			OMD	observed mean nearest neighbor distance between elements
			ENR	expected mean number of points within a circle or radius r
			ONR	observed mean number of points within a circle of radius r
			PLR	percolation from left to right of the landscape map
			PTB	percolation from top to bottom of the landscape map
			p_c	critical proportion threshold for percolation
			R	probability of finding at least one occupied cell after k moves
			R_c	critical probability threshold for finding at least one occupied cell after k moves

Proximity

PROX1	proximity index	Patch	I_{ij}	interaction between elements i and j based on a gravity model
PROX2	proximity index based on a gravity model	Patch	G	proportionality gravity coefficient
MPROX1	mean proximity index	Class	k	constant or parameter
MPROX2	mean proximity index based on a gravity model	Class	D_m	distance to mainland

Connectivity

ONL	observed number of links in the network	Network	MNL	maximum number of links in a network
V	number of vertices (nodes) in a network	Network		
ICON	index of connectivity	Network		
HI	the Harary index of landscape connectivity	Network		
IIC	integral index of connectivity of the landscape	Network	MAX	maximum value of the landscape attribute of interest (e.g., total area of the landscape or total area of a certain patch type)
dIIC	connectivity value of the patch, relative difference of connectivity after removal of the patch	Patch	IIC_{after}	integral index of the connectivity of the landscape after removal of a patch

General diversity

SH	the original Shannon index	Species, landscapes		
SI	the original Simpson index	Species, landscapes		
BP	the Berger-Parker index	Species, landscapes		
H_a	"diversity numbers" of the Hill series (equivalent number of classes)	Species, landscapes		
E_a	evenness measures related to the values of the Hill series	Species, landscapes		

(Continued)

Metrics		Application context	Related variables	
Species diversity				
p_j	proportion of individuals of species j in the total number of individuals of all species	Species		
SDH_k	species diversity of habitat k	Class		
z	slope of the log–log species–area plot	Class		
ASDH	average species diversity of a habitat (average for all habitats)	Landscape		
SDL	species diversity of a landscape	Landscape		
Landscape diversity				
p_j	proportion of class j in the landscape	Class		
HDL	habitat diversity of a landscape	Landscape		
WHDL	weighted habitat diversity of the landscape	Landscape	w_k	weight or value of habitat k
Habitat use and selection				
$HUDS_j$	habitat use diversity of species j	Species	u_{kj}	proportion of habitat k in the distribution of species j
GHUD	global habitat use diversity (for all species)	Species		
AHUD	average habitat use diversity of a species (average for all species)	Species		
SR_k	selection ratio of habitat k	Class		
IE_k	index of electivity of habitat k	Class		
Landscape configuration				
ADL	diversity of adjacencies in the landscape	Landscape	ad_{jk}	proportion of adjacencies between patch types j and k
AEL	adjacency evenness of the landscape	Landscape		

IP	proportion of interior (like) adjacencies	Landscape		
EP	proportion of exterior (edge) adjacencies	Landscape		
IEDI	interior/edge diversity index	Landscape		
IDL	diversity of interior (like) adjacencies in the landscape	Landscape		
EDL	diversity of exterior (edge) adjacencies in the landscape	Landscape		
EEL	edge evenness of the landscape	Landscape		
EDCON	edge contrast index for the landscape	Landscape	CON_{jk}	contrast between patch types j and k (weight)
IJI	interspersion and juxtaposition index (a measure of edge evenness)	Landscape		

Landscape dynamics

V(t)	vector of landscape composition at time t	Landscape	$p_j(t)$	proportion of class j in time t
AT	absolute transition matrix	Landscape	at_{jk}	elements of the absolute transition matrix indicating the area (or number of cells) that changed from state j into state k
RP	relative proportion matrix	Landscape	rp_{jk}	relative proportion of the area of the landscape that was in class j and changed to class k
M	relative transition matrix	Landscape	m_{jk}	probability that a cell (or patch) in class j changes to class k
EQ	equilibrium vector representing the composition of the equilibrium landscape	Landscape	eq_j	proportion of class j in the equilibrium landscape
CI	coefficient of inertia (or agreement) between two maps	Landscape		
CA	coefficient of agreement (or inertia) between two maps (Cohen's kappa)	Landscape	t_d	disturbance interval (time between two consecutive disturbances)

(Continued)

Metrics		Application context	Related variables	
Landscape dynamics				
TDL	transition diversity of the landscape	Landscape	t_r	recovery time (time required for recovery after disturbance)
ICT	index of complexity of temporal changes (evenness of transitions)	Landscape	d	probability that a disturbance event occurs changing from mature to initial (seral) stages of the succession
DSI	disturbance/stability index	Landscape	d_j	disturbance probability for class j
			e	probability that a cell (or patch) changes state with ecological succession to more mature stages
			e_j	probability that a cell (or patch) in class j changes for the next state with ecological succession
			M'	transpose of the relative transition matrix
			RP'	relative proportion matrix of two random maps with given compositions
			rp'_{jk}	elements of the relative proportion matrix of two random maps showing the relative proportional changes between classes j and k
			CI'	coefficient of inertia between two random maps with given compositions
			w_v	weight (proportion) of the condition v in the transition process
			m_{jkv}	probability that, under condition v, a cell (or patch) in class j changes to class k

Index

NOTE: Bold page numbers denote pages where definitions and major discussions of concepts or metrics occur.

Applied Landscape Ecology, First Edition. Francisco Castro Rego, Stephen C. Bunting, Eva Kristina Strand and Paulo Godinho-Ferreira.
© 2019 John Wiley & Sons Ltd. Published 2019 by John Wiley & Sons Ltd.